Wildlife Radio Tagging

Biological Techniques Series

J. E. TREHERNE
Department of Zoology
University of Cambridge
England

P. H. RUBERY
Department of Biochemistry
University of Cambridge
England

Ion-sensitive Intracellular Microelectrodes, *R. C. Thomas,* 1978
Time-lapse Cinemicroscopy, *P. N. Riddle,* 1979
Immunochemical Methods in the Biological Sciences: Enzymes and Proteins, *R. J. Mayer* and *J. H. Walker,* 1980
Microclimate Measurement for Ecologists, *D. M. Unwin,* 1980
Whole-body Autoradiography, *C. G. Curtis, S. A. M. Cross, R. J. McCulloch* and *G. M. Powell,* 1981
Microelectrode Methods for Intracellular Recording and Ionophoresis, *R. D. Purves,* 1981.
Red Cell Membranes—A Methodological Approach, *J. C. Ellory* and *J. D. Young,* 1982
Techniques of Flavonoid Identification, *K. R. Markham,* 1982
Techniques of Calcium Research, *M. V. Thomas,* 1982
Isolation of Membranes and Organelles from Plant Cells, *J. L. Hall* and *A. L. Moore,* 1983
Intracellular Staining of Mammalian Neurones, *A. G. Brown* and *R. E. W. Fyffe,* 1984
Techniques in Photomorphogenesis, *H. Smith* and *M. G. Holmes,* 1984
Wildlife Radio Tagging, *R. Kenward,* 1987
Principles and practice of plant hormone analysis, Volumes 1 and 2, *A. Crozier* and *L. Rivier,* 1987
Immunochemical Methods in Cell and Molecular Biology, *R. J. Mayer* and *J. H. Walker,* 1987

Wildlife Radio Tagging

Equipment, Field Techniques and Data Analysis

Robert Kenward

Institute of Terrestrial Ecology
Furzebrook Research Station
Wareham, Dorset

ACADEMIC PRESS
Harcourt Brace Jovanovich, Publishers
London   San Diego   New York
Boston   Sydney   Tokyo   Toronto

COPYRIGHT © 1987, BY ACADEMIC PRESS LIMITED
ALL RIGHTS RESERVED.
NO PART OF THIS PUBLICATION MAY BE REPRODUCED OR TRANSMITTED IN ANY FORM
OR BY ANY MEANS, ELECTRONIC OR MECHANICAL, INCLUDING PHOTOCOPY, RECORDING,
OR ANY INFORMATION STORAGE AND RETRIEVAL SYSTEM, WITHOUT PERMISSION IN
WRITING FROM THE PUBLISHER.

ACADEMIC PRESS LIMITED
24–28 Oval Road,
London NW1 7DX

United States Edition published by
ACADEMIC PRESS INC.
San Diego, CA 92101

Third printing 1993

British Library Cataloguing in Publication Data
Kenward, R. E.
  Wildlife radio tagging: equipment, field techniques and data analysis.—
(Biological techniques series).
  1. Telemeter (Physiological apparatus)
  2. Electronics in biology   3. Tracking and trailing
  I. Title   II.Series
  591.1'028   QP55

ISBN 0-12-404240-6

Typeset by Textflow Services Ltd., Belfast
Printed in Great Britain at the University Press, Cambridge

Preface

Biologists use animal radio tags for two main purposes: to locate study animals in the field, and to transmit information about the physiology or behaviour of wild or captive animals. These uses can be described, respectively, as "radio tracking" and "radio telemetry", the latter term being derived from the Greek words for distance and measurement.

Wildlife radio tags were first used for telemetry. One of the earliest projects, inspired by the use of physiological telemetry on U.S. Navy test pilots, resulted in an implanted transmitter to monitor chipmunk heart-rates (Le Munyan *et al.*, 1959). This study closely coincided with a publication from Norway describing an externally-mounted transmitter for telemetering heart and wing beats from mallard (Eliassen, 1960). The construction of these first tags, in the late 1950s, was crucially dependent on the development of the transistor.

The first radio tags transmitted a continuous signal, with physiological changes being indicated by slight changes in the signal frequency. Similar frequency modulation (FM) of a continuous carrier signal is still widely used for medical telemetry of animals in laboratories, because it is a very accurate way of conveying subtle changes in muscle potentials, neural activity, joint pressures and other physiological parameters.

For wildlife radio tags, however, the cell life can be greatly extended if the signal is transmitted as brief pulses of the carrier frequency. In theory, at least, one 25 ms pulse could be repeated every second for 40 times as long as a continuous signal from the same cell. Moreover, a faint pulsed signal is easier for the human ear to detect, against the continous background noise from wideband FM broadcasts or cosmic radiation, than a continuous whine. Wildlife radio tracking has therefore, since its start in the early 1960s (e.g. Cochran and Lord, 1963; Marshall and Kupa, 1963), been based almost exclusively on tags built for pulsed signals.

During the last two decades, radio tagging has become an important biological technique. A great deal of useful information has been published on the subject, in the proceedings from conferences in Europe (e.g. Amlaner and MacDonald, 1980; Cheeseman and Mitson, 1982), from the proceedings of the International Conferences on Wildlife Biotelemetry held in North America, in technical notes from research organizations or equipment suppliers, and in hundreds of other scientific papers. Indeed, there is so much published material that the beginner hardly knows where to start looking for simple advice. Even those with experience have difficulty deciding which equipment to use, or how best to collect and analyse their data.

This book is a general guide to radio tracking and activity monitoring with pulsed-signal radio tags. The most elementary tags have constant pulse rates and are used for radio tracking, either to sample an animal's position with single fixes or for radio surveillance: using the radio tag to find the animal so that it can be watched, captured or monitored in other ways. Such tags can also have their pulses modulated by a variety of simple sensor subcircuits to telemeter temperature, posture, movement, compass orientation and other aspects of animal activity. These tags are normally worn externally, to simplify replacement and to obtain the strongest signals. I have included a brief review of implantation, which is the best way to tag some species and is now the rule for physiological telemetry. Implanting has not been licensed for wildlife research in the United Kingdom, but is permitted under the new Animals (Scientific Procedures) Act 1986. Biomedical tagging, using frequency modulation or sophisticated pulse triggering, is beyond the scope of this book.

I have tried to write for wildlife biologists at several levels of experience. The first chapter is set at the most basic level, for those who are wondering whether radio tagging might help in their research. It compares radio tagging with other marking techniques, describes the weights, reception ranges and operating lives to be expected of radio tags, and outlines the conditions which are necessary for successful projects.

Those at a more advanced stage in their project planning will find advice on choosing receiving equipment and tags in Chapter 2. Chapters 3 and 4 provide a "branch line", for biologists who have enough electronics experience or technical assistance to consider making their own tags. Chapter 5 continues the main theme, by discussing a wide variety of tag mounting techniques. Chapter 6 describes the principles and techniques used for radio tracking on foot, or in cars, boats and aircraft: those who wish to use fixed stations, for radio location or for logging activity data, should consult Chapter 7. The final chapter provides some advice on analysing data from radio tracking and telemetric studies, and introduces a new range analysis technique.

On the whole, I have followed a sequence which is aimed at the first-time user, who need to know about the controls on receivers before using them in the field, and who might not understand a discussion of data analysis before learning about tracking techniques. The reader who has already done some radio tagging, however, may wish to start well into the book, to learn about the types of tag which suit a new study species (Chapter 5), or as an introduction to automatic data logging (Chapter 7). Moreover, although the chapter on data analysis is the last, it should be read before starting fieldwork, or its suggestions and recommendation may come too late.

Many friends and colleagues have helped in the preparation of this book.

Tom Dunstan and Mark Fuller gave invaluable advice when I started radio tracking. My knowledge of transmitter construction owes much to Mike Dolan, who first showed me how to build them, to John French who stressed the need for careful tuning, and to Brian Cresswell who has improved my understanding of circuit function and helped to develop new tags. Others who have given technical advice include Bill Cochran, Kevin Nicholas, Ray Rafarel, Trevor Storeton-West, Colin Sunderland, and engineers from Gould-Activair, Saft (UK) Ltd, IQD, and Hy-Q international. As a biologist, I would have been lost without all their help. Comments leading to tag and receiver improvements have been provided especially by Frank Doyle, Jessica Holm, Mats Karlbom and Tim Parish, together with many Biotrack customers, while Vicky Meretsky, Bruce Don, Ralph Clarke and Steve Tapper have aided my understanding of analysis techniques. Vital support for my fieldwork has been given by Jack Dempster, David Jenkins, Vidar Marcström, Mike Morris, Ian Newton and Chris Perrins, not to mention my very patient wife Bridget. Manuscript drafts were substantially improved by Brian Cresswell, Mike Fenn, Ron Mitson and David Macdonald, with the figures prepared by Jessica Holm and Stephen Hayward. I am deeply grateful to all these people.

Contents

Preface v

1
First Questions
   I. Is Radio Tagging the Best Approach? 1
  II. Do I Have Time for Radio Tagging? 2
 III. Can I Catch Enough Animals? 3
 IV. Can They Be Radio Tagged? 3
   V. Can I Afford It? 7
 VI. Can I Collect Appropriate Data? 8

2
Basic Equipment
    I. Frequency 9
   II. Receivers 11
  III. Receiving Antennas 15
  IV. Transmitters 22
   V. Tag Design 28
  VI. Tags for Tracking by Satellite 44
 VII. Tags as Capture Aids 45

3
Building Transmitters
   I. Tools 48
  II. Single-stage Transmitters 49
 III. A Two-stage Transmitter 62

4
Tag Assembly
   I. Power Sources 67
  II. Antennas 70
 III. Circuit Completion 77
 IV. Tag Designs 83

5
Tag Attachment
  I. Avoiding Adverse Effects 99
 II. Sedation During Tagging 101
III. Attachment Techniques 102
 IV. Tag Adjustment and Detachment 111

6
Radio Tracking
  I. First Principles 115
 II. Making a Start 116
III. Practice Tracking 126
 IV. Signals from Tagged Animals 130
  V. Motorized Tracking 134
 VI. Data Recording 143

7
Fixed Stations
  I. Radio Location 151
 II. Presence/Absence Recording 156
III. Radio Telemetry 159

8
Some Analysis Techniques
  I. Density Estimates 163
 II. Survival Estimates 165
III. Event Records 167
 IV. Distance and Speed Records 168
  V. Range Analysis 168

References 179

Appendix I: Addresses of equipment suppliers 193

Appendix II: Addresses of component suppliers . . . 195

Appendix III: BBC BASIC programs for range analysis 199

Index 217

1
First Questions

In an age of electronic gadgets, radio tagging is a seductive technique. For shy or elusive animals, and those too small to carry conspicuous colour tags or other visible markers, radio tracking is often the only way to collect data systematically on behaviour and some aspects of demography. Moreover, automatic radio telemetric recording can be used to record data from large numbers of unseen animals 24 hours a day.

Nevertheless, much research effort and money has been wasted on poorly planned radio tagging. Before committing scarce time and funds to the purchase of expensive equipment, it is well worth asking a number of questions.

I. IS RADIO TAGGING THE BEST APPROACH?

There is much satisfaction to be gained by radio tagging a relatively unstudied species. Tracking a few animals and thus gaining an intimate picture of their habits can be great fun, but it takes a great deal of time. Your best approach might be to put visual markers on hundreds of animals, instead of radio tagging a dozen, or to analyse contents from hundreds of stomachs instead of watching a few individuals feed. You may need to tag a great many animals in order to make valid comparisons between sexes, age, classes, areas and experimental treatments.

It is worth considering the relative advantages and disadvantages of some alternatives to radio tagging. Visual marker tags, including Beta lights and other light-emitters at night (Macdonald, 1978; Buchler, 1976; Wolcott, 1980a), are better than radio tags for identifying individual animals among others in a group, and are often adequate for making systematic observations of animals which live in the open. There are good reviews of visual markers in Stonehouse (1978) or Bub and Oelke (1980). In dense cover, on the other hand, animals with small ranges can be tracked intensively for short periods by attaching spools of very long, very fine thread (Miles *et al.*, 1981). Fastening on vegetation along the animal's path, this thread leaves a three-dimensional movement record which is much more accurate than could be obtained by radio tracking. Similarly, some mammals can be given food containing indigestible coloured fibres or beads (Randolf, 1977; Kruuk, 1978), or injected with a low activity radioisotope (Kruuk *et al.*, 1980; Jenkins, 1980), so that their faeces (and scent-marking sites?) can be identified as an indicator of their territory boundaries or dispersal movements.

Radioactive tags are a lightweight, long-lasting alternative to radio tags for animals with very small ranges (Bailey *et al.*, 1973; Linn 1978). A grid of probes to log the movement of these tags (Airoldi, 1979; Ricci and Vogel, 1984) can be built for £2000–4000, although the technique identifies only one or two individuals (with different strength radiation sources) in the same area at a time. Long life at low range can also be obtained with passive transponders, which emit a weak resonant signal if they are activated by a powerful nearby transmitter on the correct frequency. The most sophisticated transponders, measuring 2×10 mm and available from Fish Eagle Trading Company (see Appendix I for address), return a coded signal for individual identification at up to about 30 cm in water. Longer ranges, of about 1 m per 10 mm of antenna, are available from similar-sized diode transponders which were developed to find avalanche victims (Recco HB), but the necessary transceiving system works for only 1.5 h on a full charge, weighs 15 kg and costs about £10000.

II. DO I HAVE TIME FOR RADIO TAGGING?

It is very easy to underestimate the delays which can beset a radio tagging project. For a start, it will take time to select and obtain appropriate equipment. This need be no great problem in simple radio tracking projects, provided that adequate funds are available. Advice on how to select equipment which gives adequate detection range, life, reliability and attachment requirements can be obtained from other users, from equipment manufacturers and in Chapter 2. For projects at the boundaries of the technology, however, or where shortage of funds necessitates a "do-it-yourself" approach, development delays are often much longer than expected.

The final straw for many projects has been the failure to appreciate how long it can take to develop field techniques. For instance, you may need to learn how best to approach tagged animals without disturbing them, or how to interpret their behaviour from radio signal characteristics. It may take several months to discover that it is impossible to collect data as you intended, and more time to develop another way of gathering the information. By that time your sample of tagged animals may have become too small, because of deaths, dispersal and radio failures, with further delay before you can trap and tag some more. Experienced researchers often obtain few useful data in the first year of radio tagging a new species, so it is certainly no technique to grasp at in the last year of a tricky doctoral thesis.

III. CAN I CATCH ENOUGH ANIMALS?

There is no point in obtaining large quantities of expensive radio equipment before you can catch your animals. Even the well-proven techniques often seem easier when first described than when you are trying them out in the field. It is not just a matter of how the trap works, but of where best to place it, how best to bait it, how often to visit it, and the little bit of stick that needs to be wedged in just the right place to stop the wind setting it off! If at all possible, get someone who has used the trap to demonstrate it for you. You may need to go on a proper instruction course, for example if you are going to use mist nets, or anaesthetic darts which contain dangerous drugs (you can radio-track the darts too, to find the sedated animal or shots that miss). Consider buying only the minimum radio equipment, or borrowing at least the receiving system, until you are sure that you can catch enough animals for your study.

Make sure, too, that your capture technique is not unacceptably biased. If you wish to tag adult animals, you could waste a lot of time finding that only the inexperienced juveniles readily enter the traps. Bear in mind that trapping may select the hungrier or more active segments of the population. This is an important consideration if you need representative data on mortality or emigration rates. Moreover, take care that neither the capture technique nor the tags themselves unduly shock the animals, or cause other adverse effects (see Chapter 5).

In a predation study, where there is the possibility of tagging either a predator or its prey, ease of capture may affect which species is tagged. Tagging the predator can provide a variety of data on its ecology and diet, needs relatively few tags, and aids the recovery of fresh kills for analysis of selection effects, but requires additional data on predator and prey densities for estimating the predation impact (Kenward, 1980). If you want to tag the prey, you must be able to catch and tag large numbers of them, and there may be difficulty attributing deaths to a particular predator, but you can gather data on general ecology of the prey and on mortality due to all causes.

IV. CAN THEY BE RADIO TAGGED?

Two important constraints to radio tagging are the signal propagation conditions and the animal's size. For instance, the propagation of radio waves through water depends strongly on its conductivity (Fig. 1.1). With water of low conductivity, such as in rivers and freshwater lakes with a conductivity less than $0.01\ \text{S m}^{-1}$ (mho m^{-1}), the surface range of a radio tag may be reduced by only 50% at a depth of 10 m. However, the range drops sharply with depth as conductivity increases. Acoustic tagging (sonar) is therefore used when conductivity is greater than about $0.5\ \text{S m}^{-1}$ (Keuchle, 1982), in all brackish and salt water (recent reviews in Stasko and Pincock, 1977; Ireland and Kanwisher, 1978; Priede, 1980; Harden Jones and Arnold, 1982). On the other hand, radio tracking is preferable in rivers, unless they are very large and dirty. It avoids the inconvenience of having to immerse hydrophones in the water; moreover, the detection range of acoustic tags is low in turbulent and aerated water, especially if it is shallow. Migrating fish can be radio located with rapid searches along riverside roads or from aircraft.

Most invertebrates and some terrestrial vertebrates are too small to be radio tagged. The smallest crystal-controlled VHF tags weigh about 800 mg,

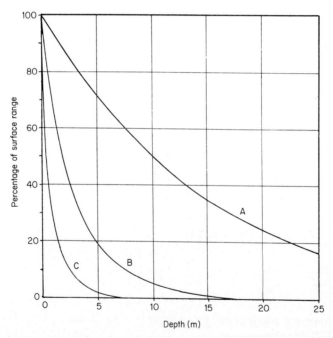

Fig. 1.1. The surface range of radio tags in fresh water reduces as their depth increases. The curves show theoretical range loss as conductivity increases from $7 \times 10^{-3}\ \text{S m}^{-1}$ (*A*), through $30 \times 10^{-3}\ \text{S m}^{-1}$ (*B*) to $80 \times 10^{-3}\ \text{S m}^{-1}$ (*C*). Redrawn, with permission, from Keuchle (1982).

and can be used for animals down to about 8 g in weight. The smallest birds and bats seem able to fly well with packages at 10–15% of their weight (Graber and Wunderle, 1966; Stebbings, 1982; 1986), although a 4–6% loading increase is probably enough for animals above 50 g (Brander and Cochran, 1971; Cochran, 1980), and 2–3% is a wiser limit for the largest birds and mammals.

Tags which lack crystal control can be built with a total weight of 700 mg (Mohus, 1983) or less. However, these tags radiate their power over a much broader frequency band than tags with crystal control, which substantially reduces efficiency and hence detection range, and hinders individual identification. Such tags are also illegal in many countries (including the United Kingdom).

There is little prospect of appreciably smaller tags in the immediate future, because the smallest power cells and VHF crystals each weigh about 300 mg. One must allow 200–400 mg for other components, antenna and potting materials. Expansion of thin-film oscillator technology may eventually make it possible to get reasonable detection ranges from even smaller tags, but perhaps only if circuits suitable for wildife tags are developed for other, more extensive commercial use (just as the smallest cells are made primarily for the hearing-aid and wrist-watch markets).

Tags for small animals also have limited detection ranges and operating lives. The smallest crystal-controlled tags, containing 300 mg mercury or silver-oxide cells, can transmit for 10–15 days and give maximum line-of-sight ranges at ground level of from 100 m, with integral tuned-loop antennas, to 3 km with 25 cm whip antennas of 0.20 mm (0.010 in) guitar wire. A 1.3 g package with a 500–600 mg cell gives a more useful 3–5 week life, with 3–5 months possible from a 4 g tag containing 2 g cells, and 10–20 months from 12 g tags containing 7 g lithium cells. All these packages contain simple single transistor ("single-stage") oscillator circuits, and their maximum detection range depends mainly on their antenna dimensions. Using the most sensitive hand-held receiving equipment, line-of-sight ranges of more than 20 km are possible, even to single-stage tags (with 30 cm whips and 20 cm "ground-plane" antennas on flying birds, although the range may be less than 1 km to such a bird on the ground in woodland). The importance of using long whip antennas cannot be stressed too strongly, and it is also worth noting that a thick antenna radiates somewhat better than a thin one (Fig. 1.2).

Longer ranges can be obtained with two-stage transmitters, which contain extra circuitry to amplify the oscillator output. The smallest two-stage transmitters, using thick-film and surface-mount technology, weigh 1–1.5 g. Such tags can give 100% greater ranges than single-stage tags of the same size. Larger two-stage tags give two to three times the range of the best

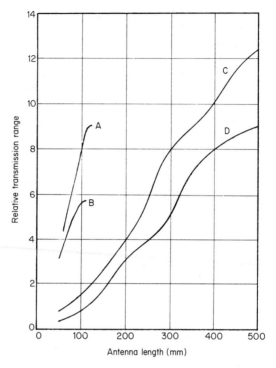

Fig. 1.2. The range of radio tags increases with antenna dimensions. Moreover, at a given circumference, tuned loops made from 8 mm brass strip (*A*) gave longer ranges than loops of 1 mm multistrand copper wire (*B*), and whips of 2 mm steel cable (*C*) were more efficient radiators than whips of 0.2 mm nickel guitar wire (*D*). The range curves are estimated, assuming no propagation losses, from the power output of a 150 MHz transmitter which was tuned to each antenna. Antennas were orientated for peak signal transmission, and were in air: ranges are 30–50% less from loops on animals.

single-stage types, and even more powerful tags are built, for example for tracking by satellite. Nevertheless, there is a penalty to be paid for the extra range: two-stage tags usually cost 30–60% more than single-stage equivalents, and have shorter cell-lives. To transmit twice as far theoretically requires four times the signal output (by the inverse square law), so tags with the same conversion efficiency (electrical energy to radio signal) theoretically require four times as much cell capacity to double their range. In reality, two-stage tags tend to have higher conversion efficiencies than single-stage designs, so that doubling the range may reduce life by only 40–50% at the same tag weight.

All this means that some projects are impractical simply because the animals cannot carry tags which are powerful enough, or because tag lives would be too short if the power output was adequate for tracking. This is a common problem for small animals that have very large foraging ranges (e.g. some bats and small birds), but can also affect larger species whose tags cannot be equipped with efficient antennas (e.g. implanted or ingested tags for amphibia and snakes). It is not yet possible to get 5 km range across flat ground from tags swallowed by a 100 g toad.

V. CAN I AFFORD IT?

When research funds are short, it is tempting to try and save money by building your own equipment. This is generally unwise for the newcomer to radio tagging, but can sometimes be worth while.

With a variety of reliable radio tracking and telemetry receivers now available commercially (see Appendix I for addresses), the cost in time of developing one's own is likely to be much greater than the cost of purchase. A possible exception is when detection ranges need be no more than 1–200 m from transmitters which radiate strongly, in which case FM/AM broadcast receivers with frequency converters have sometimes proved adequate. For instance, an inexpensive Sony 5900W receiver has been used, with a frequency converter, for tracking hedgehogs equipped with 104 MHz tags (Morris, 1980). Nevertheless, for animals smaller or faster-moving than hedgehogs, or as an investment for future projects which may embrace such species, it makes more sense to buy a proper radio tag receiver. This will cost at least £250, for a lightweight receiver with a limited frequency band, or upwards of £550 for a robust receiver to use with at least 30 different tag frequencies.

Radio tags too can be bought from a number of firms, as ready-to-mount packages or as basic transmitters to which power cells and antennas must be added. Ready to mount, single-stage tags cost £40–£90, or £55–£140 for two-stage types. Financial savings of 20–50% are possible by adding cells, antennas and mounting materials to basic commercial transmitters. This approach, covered in Chapter 4, also gives some flexibility in package design. However, it is important to realize that design changes may have a considerable effect on the tuning, and hence the signal output. Moreover, cells and the transmitters themselves can be damaged by careless soldering, while clumsy potting can add unnecessary weight or allow moisture to penetrate. Unless you have some experience, and are using a simple tag design, you may be wise to rely on the manufacturer's packaging skills.

Even more money can be saved by making one's own transmitters, as described in Chapter 3. Given patience, reasonable manual dexterity, and the right tools, most people can get some sort of signal from a simple oscillator circuit, especially transmitter designs on a printed circuit board

(PCB). However, to make well-tuned and reliable tags requires practice, which may take many weeks in the case of the smallest single-stage transmitters. Moreover, building radio transmitters requires a testing and development licence in the United Kingdom and some other countries, and there may be problems in obtaining components, due to large minumum-order charges or obsolescence. Bearing in mind the cost of tools, and the time that must be spent in learning construction skills (and searching for animals whose tags failed prematurely!), it probably is not worth learning to build single-stage transmitters unless you will need several hundred of them, or more than about 50 of the PCB-based two-stage transmitters.

VI. CAN I COLLECT APPROPRIATE DATA?

Having avoided the problem of collecting too few data, some projects have collected plenty, and then been baulked by a lack of satisfactory analysis techniques. For instance, long strings of consecutive observations tend to lack statistical independence, or data may be recorded automatically in such large quantities that local computing facilities are overloaded.

Effective project planning should consider all stages of the study, right through to data analysis. Even if you have satisfactory answers to all the previous questions, no definite decision should be made until you have considered that final stage, with help if necessary from other users, one or more statisticians, and the literature (make a start with Chapters 6 and 8). If you plan your data collection to simplify the analysis, buy the right equipment for collecting that data, and allow plenty of time for teething problems, you will have given yourself the best chance of success.

2
Basic Equipment

This chapter provides a review of the receiving and transmitting equipment used in wildlife radio tracking and radio telemetry. It concentrates on items which are commercially available or reasonably easy to build, but also mentions advanced developments which may be used more widely in the future, giving references for those who need further information on a particular topic. After this chapter, those with no interest in building their own tags should jump to Chapter 5.

I. FREQUENCY

Many countries now allocate a frequency band of 150 kHz–2 MHz for wildlife radio tags, mostly in the Land Mobile Band at 138–174 MHz. The United States also has bands at 40 MHz and 216 MHz (Kolz, 1983), while biologists in Finland use 230 MHz. Even higher frequencies are used for some specialized applications, including tags for tracking from satellites.

The United Kingdom originally had a protected frequency allocation for radio tracking at 102 MHz, but this was later shifted to 104.6–105.0 MHz. Another band was available at 173.20–173.35 MHz. Unfortunately for many biologists, the lower band is being lost to radio broadcasting, which increases its allocation up to 108 MHz between 1985 and 1995, following a decision at the 1979 World Administrative Radio Conference. In compensation, a further band has been allocated at 173.70–174.00 MHz. Biologists in some countries have access to a band at 27 MHz, but this is now relatively seldom used, partly because of interference from Citizens Band (CB) radio and partly because small antennas are more efficient at the higher frequencies (see below).

Since there is continual growth in the number of radio applications competing for frequencies, radio tags often share their allocation with other low-power transmitters. This is the case with the British 173.20–173.35 MHz band, which is shared with narrow-band industrial telemetry and telecontrol equipment. Such sharing has been satisfactory for wildlife biologists who work only in rural areas, but the band is becoming increasingly crowded, especially with security devices, in built-up areas. Where another transmission is strong in their area, biologists must avoid having tags on the same 3–5 kHz as the other narrow-band transmission. Although the 173.70–174.00 MHz band is also shared, with users of radio-microphones, interference is at present far less likely from these than from industrial users of 173.20–173.35 MHz (although there may be some noise close to 174 MHz from Band III television in neighbouring countries). Animal tags are usually of such low power that interference caused to other users of the same frequency band is extremely unlikely.

Where there is a choice of frequencies, several factors must be taken into account, including antenna efficiency and directional accuracy, habitat effects on signal propagation, and ease of transmitter construction. Antenna dimensions are often expressed as fractions of a wavelength. These may be readily calculated from frequency, since radio waves travel at about 300×10^6 m s^{-1}, by the formula:

$$\text{wavelength } (\lambda, \text{ in m}) = \frac{300}{\text{frequency } (f, \text{ in MHz})}$$

For instance, at 150 MHz the wavelength is 2 m. A simple receiving antenna for 150 MHz is a $\lambda/2$ dipole, but the most common hand-held receiving antenna at this frequency is a three-element Yagi antenna (Fig. 2.1), with a longest element length of about 1 m. This antenna, unjustly called after the translator of the original Japanese design, by Uda and Mushiake (1954), has about twice the gain of the dipole and is far better for taking bearings.

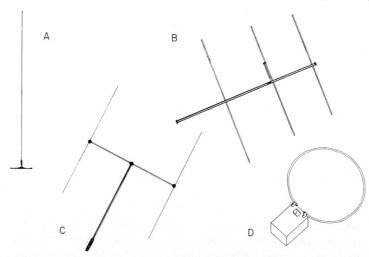

Fig. 2.1. The antennas most commonly used for mobile radio tracking: a dipole for car roof mounting (*A*), a three-element Yagi (*B*), an H-Adcock (*C*) and a loop antenna mounted on a receiver (*D*).

The longest element of a $\lambda/2$ Yagi at 104 MHz is nearly 1.5 m, which is too long for easy handling except in the most open country. Dipoles are therefore often used for this frequency in wooded areas, with resulting loss in range and precision. In more open country, however, various H-shaped antenna designs (Fig. 2.1) can be used to give greater reception ranges and bearing accuracy. At 27 MHz a $\lambda/2$ dipole would be about 5.6 m long, as would the longest elements of a $\lambda/2$ Yagi or H antenna! For this reason, the most common receiving antenna at 27 MHz is a $\lambda/6$ loop, 0.5 m in diameter and with about half the range of a $\lambda/2$ dipole.

Although the higher frequencies have the advantage in antenna efficiency, especially since the performance of both transmitting and receiving antennas benefits as wavelength decreases, the low frequencies propagate appreciably better through water. Tags at 104 MHz have thus been especially suitable for tracking fish in the United Kingdom. North American biologists often use 40 MHz for fish tracking.

For non-aquatic species, the disadvantage of increased signal absorption by vegetation at high frequencies is normally considered a minor inconvenience, compared with the benefits of improved antenna efficiency, at least for frequencies up to 174 MHz. Nevertheless, the lower frequencies may be worth considering in habitats where the vegetation is very dense and damp, such as rain-forests. In dry, open terrain, on the other hand, one or two projects have obtained excellent ranges with frequencies above 200 MHz, for instance when tracking condors (W. Cochran, personal communication). On the whole, however, frequencies in the 138–174 MHz Land Mobile Band are a good compromise for most work on non-aquatic animals, and are convenient frequencies for building lightweight tags with low power consumption (Chapter 3).

II. RECEIVERS

A. Receivers for small projects

Radio-tag receivers can be divided into three main categories, which are sometimes available as different models from the same supplier. The most basic category is sold primarily for locating trained raptors or dogs. These receivers, which start at about £250, are not usually as robust as the receivers built for regular research use, cover a limited frequency band of 75–200 kHz, and often lack refinements as fundamental as a recorder output or a signal strength meter. Nevertheless, these receivers can be just as sensitive as the more sophisticated models, and some are small enough to fit into a large pocket. This makes them very suitable for low-budget projects or pilot studies in which no more than 7 (with 75 kHz bandwidth) to 20 (with 200 kHz bandwidth) tag frequencies will be required, especially if the emphasis is on radio surveillance.

It is possible to put two or three tags with different signal pulse rates on each frequency, but this creates problems when trying to detect weak signals or take bearings near another tag on the same frequency. A tag frequency separation of at least 10 kHz is therefore preferred in most projects. This enables tags to be individually identified despite frequency drift: some tag frequencies shift 1–3 kHz from their intended value with changes in temperature and cell voltage, because of variation in crystals or other components. It is also possible to use a tag separation of 5 kHz, provided (i) that the receivers have high selectivity and drift no more than 1 kHz from −20°C to +40°C, (ii) that the tag crystals have high calibration accuracy with low temperature drift (expensive), and (iii) that tags are made with different pulse rates on alternate frequencies. Nevertheless, this puts a lot of pressure on the tag maker. Note that the repeated retagging of 10 small animals typically requires 15–20 tag frequencies, if exhausted tags are replaced rapidly without using the same frequency twice.

B. Receivers for large projects

If more than 15 tag frequencies may be needed, perhaps in the user's or department's next project, it is sensible to buy a receiver in the second category, built specifically for research. Four reliable and widely used receiver models are shown in Plate 1. They typically differ by no more than about 3 dB m in sensitivity, at around −145 dB m to −148 dB m (0.015–0.010 μV) for the lowest signal strength that can be heard, although some individual receivers perform worse than others of the same type. A signal strength of −148 to −150 dB m is about the faintest that can be detected against background radiation and noise from components. These receivers weigh 1–2 kg with internal batteries, which is convenient for carrying round the neck: more than 2 kg can become uncomfortable. They are all straightforward to operate, although it is easiest to find tags quickly with digital tuning; otherwise one needs to know the channel number and the fine-tune position which is equivalent to each frequency, or to set a large sweep dial with considerable accuracy. The European receivers (bottom row) have integral dial illumination for nocturnal researchers, and are considerably stronger and more weatherproof than those from the United States (top row), which must be carried in plastic bags if it rains. The M-57 is almost completely watertight and has survived being run over by a Land

Plate 1. Four receivers which are widely used for radio-tracking. The CE-12 (top left, photo from Custom Electronics) and TR-2 (top right) are made in the United States, whereas the M-57 (lower left) and the RX-81 (lower right) are European.

Rover. However, it is also the heaviest (2.2 kg), and covers only a 150 kHz band. The others can cover at least 1 MHz, albeit with the addition of extra 300 kHz bands to the basic 300 kHz bandwidth CE-12 (=LA-12). Their temperature stability varies, and is particularly bad on some old LA-12s. These may drift 5 kHz between a sunny winter afternoon at +15°C and the next chilly morning at −10°C. The tuning dial can easily be set 5 kHz lower to compensate for this during tracking work, but the signal tends to be lost if the temperature changes markedly during automatic recording.

All have panel meters to indicate signal strength and the voltage in their 9–12 V supplies, which can be primary cells, rechargeable NiCads, or an external source such a vehicle battery. All have integral loudspeakers, as well as jackplug sockets for earphones or earplugs, and separate sockets for recorder plugs. A separate gain and volume control can be useful for automatic recording, and is standard on the M-57 and RX-81. The latter also has a separate automatic gain control (AGC) switch. A sweep control, which enables 5–10 kHz to be swept automatically for tags whose precise frequency is unknown, is standard on the RX-81 and optional on the other models.

These research receivers are available from around £360 (1986 prices), for the Mariner Radar M-57 (150 kHz), up to $1400 for the Telonics TR-2 model for 216–218 MHz. The mainstay of projects in the 1970s, the LA-12, is available as the CE-12 from Customs Electronics and as the RB4 series from Telemetry Systems for from $625 (300 kHz) to $725 (1.5 MHz), but no longer from AVM, who sold it originally. The RX-81 (1 MHz) costs about £600 from Televilt or Biotrack. Since the receivers are so similar in performance and reliability, most users base their choice as much on price (including possible shipping and import charges), familiarity and ease of servicing as on real differences between the models.

C. Programmable receivers

The third category of receiver can be programmed, from an integral or external memory, to scan through a number of tag frequencies. This facility is useful mainly (i) when searching for a number of lost animals, and (ii) for automatic data recording from many tags at one site.

In projects where several animals at a time may go out of range from normal tracking points, for instance through dispersal movements or being carried away by predators, it becomes very tiresome to set tuning controls repeatedly for each tag frequency, especially when driving a car or moving fast across country in an aircraft. A programmable receiver can be set to tune each frequency automatically, usually for 3–10 s, in a repeating sequence which can be stopped when a signal is detected. This saves wear on the tuning controls and leaves the searcher's hands and mind free to concentrate on getting from place to place.

Since non-programmable receivers must be tuned by hand to each tag frequency, they are normally used for automatic recording of signals from only one animal at a time. Although the presence or absence of several tags on the same frequency can be determined by listening through tape-recorded audio output, provided that each tag has a different pulse rate, this is a tedious process. A programmable receiver can be used to cycle repeatedly through a number of tag frequencies, recording for the same preset sample period on each. The signals may be output to a chart or tape recorder, or to more sophisticated data-logging equipment (see Chapter 7).

At least two programmable receivers are available. One, designed at the Cedar Creek Bioelectronics Laboratory in the University of Minnesota and sold by Advanced Telemetry Systems for $1800, has integral memory circuitry and scan-time selector. For use with a data logger, it should be obtained with optional sockets for external control of switching-on and channel changing, and for output of signal cues and channel number. This receiver can also be used for tracking on foot, but early versions were quite heavy (2.5 kg), and more easily damaged than less complex receivers. It is best kept for fast searches and automatic recording. The Telonics TR-2 receiver, costing $1250–1400, can be separated from the $800 memory and control unit, and is then suitable for normal tracking work.

Another programmable unit which has been used in some recent projects is the Yaesu FT-290R. This is a frequency-synthesized communications transceiver with 10 memory channels, which is supplied to cover 144–146 or 144–148 MHz but has also been used with a frequency converter for 104 MHz. It weighs nearly 2 kg with batteries, but this weight can be reduced by removing the transmitting section. The basic cost is about £350. Although this equipment is not weatherproofed, it can be used both for radio tracking and also as part of an inexpensive automatic recording system for small numbers of tags.

III. RECEIVING ANTENNAS

A. Portable designs

At frequencies above 140 MHz, the three-element Yagi is the most commonly used antenna for tracking on foot. Well tuned, this antenna has a gain of nearly 7 dB over the $\lambda/2$ dipole. The decibel is a logarithmic measure, and a gain of 6 dB represents a theoretical doubling of the reception range. The three-element Yagi therefore has about twice the range of the dipole.

16   Wildlife Radio Tagging

The Yagi's other big advantage is its signal reception pattern (Fig. 2.2*B*). There are peaks in signal reception along the line of the boom, with the strongest peak to the front of the antenna. Such an antenna is said to have a good "front-to-back" ratio, which makes it relatively easy to distinguish between true and reverse bearings. In contrast, the dipole has symmetrical signal peaks at right angles to its main axis, with nulls along the line of the antenna (Fig. 2.2*A*). It is therefore difficult to obtain an unambiguous bearing with a horizontal dipole (a vertical dipole is omnidirectional in the horizontal plane). Another aspect of antenna performance is the "sharpness" of the peaks and nulls in the reception pattern, which contribute to the taking of accurate bearings. With a three-element Yagi, for instance,

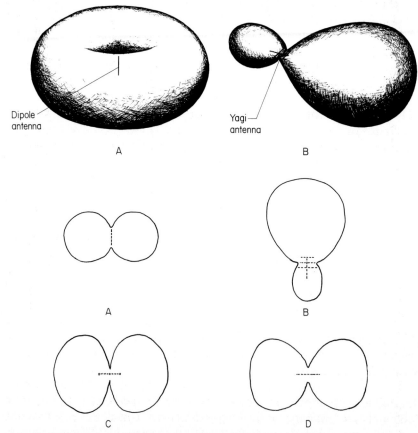

Fig. 2.2.   The radiation/reception patterns for dipole (*A*), Yagi (*B*), H-Adcock (*C*) and loop (*D*) antennas. The patterns are three-dimensional (e.g. doughnut-shaped for a dipole), but are normally shown in horizontal cross-section.

bearing accuracy in moderately open country averages about 5°, compared with 5–10° for a dipole (Amlaner, 1980; Cederlund *et al.*, 1979).

A popular portable antenna at 104 MHz is the H-Adcock design (Taylor and Lloyd, 1978). With 1.5 m vertical elements and a 1.5 m ($\lambda/2$) crosspiece at this frequency, an Adcock antenna has nearly 6 dB greater gain than a dipole (Amlaner, 1980). The signals from the two elements are fed to the receiver with a 180° phase difference (using a balun match), so the reception pattern has a sharp null when the elements are equidistant from the source (Fig. 2.2*C*). This gives the Adcock antenna an angular accuracy of about 3° (Macdonald and Amlaner, 1980), somewhat better than a three-element Yagi. However, its reception pattern is like the dipole's in producing an ambiguity between true and reverse bearings. This is usually resolved (i) by taking a cross bearing from another position, (ii) by reference to the animal's position shortly beforehand, or (iii) by reference to landscape features. For instance, the ambiguity from dipole or Adcock antennas is not very important when tracking fish from river banks.

Another type of H antenna is available from some equipment suppliers. It differs from the H-Adcock in having one element shorter than the other, with the two elements phased to give maximum signal reinforcement when the shorter element is towards the transmitter. The antenna therefore functions like a two-element Yagi, without the Adcock's ambiguity between true and reverse bearings, but with slightly less gain and accuracy than the three-element Yagi. The design's advantage over the Yagi is its smaller size, and it is probably ideal for work around 100 MHz. However, the Yagi's greater accuracy, especially in the vertical plane, is essential for precision tracking of arboreal species (i.e. when both vertical and horizontal bearings are required).

Loop antennas are preferred for tracking on foot at 27 MHz. The tuned $\lambda/6$ loop has 3 dB more gain than a (1.9 m) $\lambda/6$ dipole, with a reception pattern which gives a null at each side along the loop's axis (Fig. 2.2*D*). The $\lambda/6$ loop antenna thus resembles dipole and Adcock antennas in its tendency to give reverse bearings, with a directional accuracy which is at best about 5° (Cederlund *et al.*, 1979). Loop antennas can also be useful for short-range work at higher frequencies: at 150 MHz the $\lambda/6$ loop is small enough (20 cm diameter) to be fixed to the outside of a board like a table-tennis racket, where it is much less likely than a Yagi, H-Adcock or dipole to get snagged in dense cover. Moreover, since a very strong signal is obtained from a transmitter within the loop circumference, small open loops are handy for finding tags which have dropped off animals, for instance under muddy water (Solomon and Storeton-West, 1983).

All these antennas can be obtained at reasonable prices from manufacturers listed in Appendix I. The antenna should be supplied with a coaxial cable which connects to the receiver. Since there are two connection systems commonly used for this purpose, either with screw-on UHF connectors or with smaller, bayonet-type BNC connectors, be sure to order a cable which matches the receiver socket.

For those running projects on a shoe-string, there are instructions for building a variety of suitable types in Amlaner (1980) and in the manual of the American Radio Relay League (ARRL, 1984). Note that the gains of antennas in some texts are given with reference to a theoretical isotropic point source (dB_i), rather than with reference to a $\lambda/2$ dipole (dB_d). The former units are 2.14 dB greater than the latter ($dB_d = dB_i + 2.14$ dB).

B. Antennas on vehicles

Cars, minibuses and lorries can be equipped with bulkier but more powerful direction-finding systems than can be carried on foot (see e.g. Verts, 1963; Proud, 1969; Hallberg *et al.*, 1974; Kolz and Johnson, 1975; Cederlund and Lemnell, 1980). For instance, adding elements to a Yagi antenna increases both its gain (Fig. 2.3) and its accuracy. Thus, a twelve-element Yagi has a

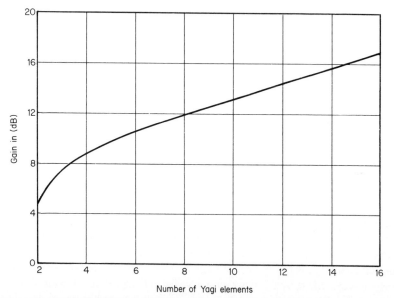

Fig. 2.3. The gain of a Yagi increases with the number of elements. The gain is given in decibels, and shows the increase with reference to a 1 dB dipole. Each gain increase of 6 dB represents a range doubling: a ten-element Yagi has about twice the range of a three-element Yagi. The gain increase from three to six elements is about 3 dB, which represents a power increase of 10^{-3}, and thus a range increase of $\sqrt{10^{-3}}$, i.e. 40%. Redrawn with permission from the American Radio Relay League.

2. Basic Equipment

gain of about 14 dB over a dipole, and some 7 dB over a three-element Yagi. This gain is not achieved without cost, however, because the higher the gain the narrower the angle (beamwidth) in which it can be obtained. The peak gain of a Yagi is obtained with the antenna pointing directly at the signal source, and this drops by 3 dB when a three-element Yagi points about 30° to either side of this line, giving a total half-power beamwidth of 60° (Fig. 2.4). The half-power beamwidth of a twelve-element Yagi is much narrower, only 30°. The narrow beam-width can be an advantage, in that it increases bearing accuracy, but can also be a nuisance: the narrower the beam, the easier it is to miss a pulsed signal if the antenna is swung round too fast. A six-element Yagi is therefore a popular compromise for vehicle-mounting (Plate 2), giving a 3 dB better gain than a three-element design, and an accuracy of about 3° (Cederlund et al., 1979).

Even greater accuracy, without loss of beamwidth, can be obtained by combining a pair of vertical 4–6 element Yagis in a "null-peak" system (Banks et al., 1975). A switch box is used to feed the signals from the two antennas exactly in phase, thus producing the optimum gain for finding the approximate direction of a tag, or out of phase to produce a very sharp null

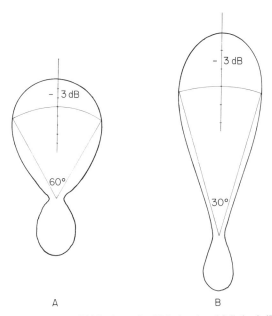

Fig. 2.4. The ideal 3 dB "beamwidth", through which the signal falls by 3 dB on each side of the peak, is 60° for a three-element Yagi antenna (*A*), compared with 30° for a 12-element antenna (*B*). To obtain ideal beamwidths, the antennas must be well tuned and at least 2λ from any surface, with their elements vertical. Figure 2.2 shows a more typical beamwidth, for a hand-held Yagi with its three elements horizontal, of about 100°.

Plate 2. A hand-raised mast mounted in a minibus. The antenna is a six-element Yagi, mounted horizontally, with a repeater compass to improve bearing accuracy.

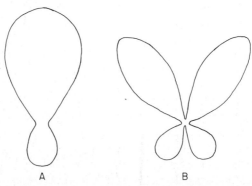

Fig. 2.5. The radiation patterns, in the horizontal plane, for twin six-element Yagi antennas with vertical elements. When the antennas are switched in phase (A), the single lobe could have a 3 dB beamwidth of about 50°, compared with a 20° beamwidth for the null when the antennas are out of phase (B).

when the array points straight at the tag (Fig. 2.5). Such systems can give a bearing accuracy of close to 1° (Hallberg et al., 1974; Cederlund et al., 1979), but tend to be too damage-prone for vehicle-mounting in rough or wooded terrain.

As well as risking damage from low branches, cables and other obstacles, vehicles equipped with directional antenna systems must usually be stopped to rotate the antenna and take bearings. So for fast-moving searches, especially in areas with a good road net or where direction-finding is difficult (e.g. towns), a simple whip is often preferred. When this omnidirectional dipole picks up a signal, bearings can be taken with a hand-held antenna. Similar restrictions may affect the choice of antenna for tracking in aircraft: a Yagi under each wing is ideal, but aircraft design, pilot preference or insurance regulations may mean that a whip is the only option. Mobile antenna systems are discussed more fully in Chapter 6.

C. Fixed stations

For non-directional recording of activity data or other telemetered information, many projects use a vertical whip with ground plane elements (Amlaner, 1980; ARRL, 1984) placed near the centre of the study area. Sometimes, however, the recording site is to one side of the study area, because of local topography or because of a convenient building in which to keep the receiving system. In this case it is often possible to use a Yagi antenna. The distance and width of the study area will determine the optimum number of Yagi elements, bearing in mind that gain increases at the expense of beamwidth. In other words, if the study area subtends an angle of about 60° at the receiving antenna, you should not use a Yagi with more than three or perhaps four elements.

On the other hand, to record the presence of a bird at a tree nest from ground level, it might be best to use a twelve-element Yagi pointed straight at the nest, with $\lambda/4$ or $\lambda/8$ dimensions if small size is more important than high gain. Another alternative would be to put a small loop antenna just under the nest. However, this could present reception problems if it required a long cable to the receiving equipment: without very effective shielding, the connecting cable would pick up signals from birds off the nest. Long cables from fixed antennas to receiving systems can also result in an appreciable signal loss. Each cable type has a figure for its loss per unit length at particular frequencies, and a preamplifier should be used if the loss will exceed about 3 dB. This is the loss in about 10 m of most coaxial antenna cables at 138–174 MHz (ARRL, 1984).

Direction-finding stations suitable for simple tags have usually been based on rotating Yagis at two or more mast sites. Automatic systems can be built, using microprocessors to assess and record the signal peaks from each site (Cochran et al., 1965; Deat et al., 1980). However, many more projects have recorded the bearings manually, either using two people at fixed sites in contact by CB radio, or a manned mast site in contact with a mobile station, or a master site which can control the antenna rotation at a slave site and monitor the signals received there (Smith and Trevor-Deutsch, 1980). Manned stations are best for detecting faint signals, for high accuracy using null-peak direction finding and for rapid rechecking of "improbable" fixes.

Other types of radio-location system can be built without rotating antennas. For instance, hyperbolic systems have been tested in the United States (Patric and Serenbetz, 1971), Australia (Yerbury, 1980) and Sweden (Lemnell et al., 1983). These systems measure minute differences in the time taken by signals to reach a master station and two or more slave sites. This requires relatively powerful tags, typically working on the same frequency but with some form of coding for individual identification, and sophisticated receiving equipment. A system is commercially available from Tracktech in Sweden (Appendix I), with a minimum tag weight of 40 g and a cost of about £25000 for a master station, two slaves and 20 tags. There may also be promise in the use of active-array antennas, which are available for small boats. Although an automatic active-array system could be built quite cheaply for use with simple tags, the active-array dipoles have less gain than a Yagi and would therefore have less range or require relatively powerful tags. Fixed station recording systems are covered more extensively in Chapter 7.

IV. TRANSMITTERS

Although the terms "transmitter" and "radio tag" are often treated as synonyms, in this book I maintain the convention of using the term "transmitter" for the electronic circuitry alone, which is combined with a power supply, antenna and mounting materials to make a "radio tag".

In the simplest crystal-controlled transmitters, signal pulses ("bleeps") are governed by a single pulse capacitor in an elementary oscillator circuit (Fig. 2.6). With all else held constant, increasing the value of this capacitor increases the pulse duration and the interval between pulses. The same form of pulse control is used in many two-stage transmitters, in which the power output is increased by a simple amplification stage (Fig. 2.7). However, the pulse characteristics in these early designs are also affected by the tuning: signal pulse rates may vary appreciably if animals ground the antenna, for

2. Basic Equipment

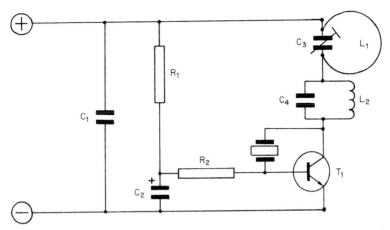

Fig. 2.6. A circuit for a "single-stage" transmitter with a loop antenna (L_1). See Chapter 3 for component values and construction details.

example by touching wet vegetation, as they move about. This is generally unimportant during tracking and activity monitoring, and can even be useful as a cue to what the animal is doing, but can be a problem in high-accuracy telemetry. More precise pulse control is possible in transmitters which contain discrete component multivibrators (Taylor and Lloyd, 1978; Thomas, 1980), or CMOS pulse control circuitry (Macdonald and Amlaner, 1980; Anderka, 1980), which may also be used to produce coded pulse trains (Lotimer, 1980), and for stability in very high power transmitter designs (Keuchle, 1982).

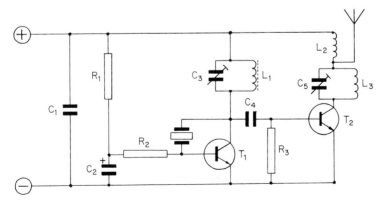

Fig. 2.7. A circuit for a "two-stage" transmitter with a whip antenna. Component values and construction details are given in Chapter 3.

24   Wildlife Radio Tagging

A. One-stage or two-stage transmitters?

This is an important question in the planning of most tagging projects. Tags built with two-stage transmitters tend to cost more than single-stage tags, and to weigh more for a given signal life, but to give a 100–200% greater range. The range increase is partly because two-stage transmitters are usually run at 2.7–3.5 V, compared with the 1.35–1.5 V typical of single-stage tags. Nevertheless, two-stage transmitters have a slight range advantage even when run at the lower voltage, because their circuits are more efficient than the usual single-stage designs.

The smallest two-stage tags, using transmitters with hybrid or surface-mount circuitry (Austek, Biotrack) and the smallest 1.35–1.5 V cells, weigh about 1.5 g. Animals too small to carry these can only be equipped with 0.8–1.3 g single-stage tags (Plate 3, top left and centre). However, single-stage transmitters are also best for very many studies of larger animals, because of their cost and life/weight ratio. For instance, the life of a 1.5 g two-stage tag, with two 24 mA h silver cells (Plate 3, top right), is at most 7–10 days and the cost at least £60. In contrast, a £45 single-stage tag of similar weight could contain an 85 mA h mercury cell and run for 4–6 weeks,

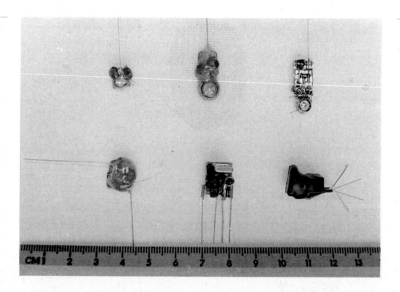

Plate 3. A selection of small radio tags and simple transmitters. The top row includes commercial single-stage tags of 0.8 g (left) and 1.3 g (middle), with a 1.5 g two-stage tag (right). Chapter 3 describes how to build the single-stage (left) and two-stage transmitters (centre, right) in the bottom row.

albeit at half the range of the two-stage design. Using two-stage transmitters which are built with discrete components, tags with 800 mA h lithium cells can weigh about 14 g and transmit for 5–8 months at 3 V. A 12 g single-stage tag with a 1.5 V lithium/copper-oxide cell would have less than half the range, but last 2–3 times as long because of increased cell capacity (1400 mA h) at the lower voltage. A two-stage tag at 3 V must weigh about 30 g to approach a two-year life, unless solar power can be used (see Section V.B).

In general, two-stage tags are best used for large, wide-ranging mammals, especially if antennas must be short and enclosed to prevent the animal damaging them, or for birds or bats which are migrating or foraging very far from their roosts. And in some cases, where single-stage tags are adequate for mobile tracking, single-stage signals may not be strong enough for automatic recording equipment, which detects weak signals less well than the human ear.

If you cannot decide in advance whether one-stage transmitters will suffice, it is worth trying these as well as the two-stage types, provided that time is available and the animals are reasonably easy to catch. It is sometimes handy to have the extra range of a more powerful tag at the start of a project. Later on, when you are familiar with the technique and the animals' habits, you can save money and gain tag life by changing to one-stage tags.

B. Buying transmitters

There are several ways of assessing the quality of commercial transmitters. Most manufacturers specify the size, weight, voltage range and current drawn by their products. The current is of course the average current through the pulsing circuit.

Some makers also stipulate the temperature drift, the pulse duration and rate, and the power output. Transmitters, like receivers, should have little temperature drift, or there may be difficulties detecting them. A drift greater than 3 kHz between +30 and −20°C, would be unsatisfactory for automatic recording.

A reduction in pulse duration or pulse rate decreases the circuit current, thereby increasing the tag life at a given battery size. Thus, a transmitter with a 20 ms pulse should last twice as long as one with a 40 ms pulse, all else being equal. However, pulse lengths should not be much less than 30 ms, at least for radio tracking rather than automatic telemetry, because the detection efficiency of the human ear falls quite sharply as pulse length is reduced below this value (Fig. 2.8). Similarly, tracking becomes difficult at pulse rates below about 50 per min. Although some biologists prefer rates of

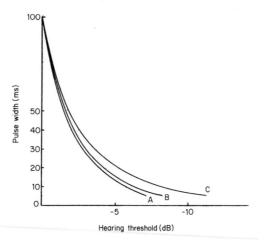

Fig. 2.8. The audibility of sound pulses declines as the pulse duration is reduced. The decline is least for high pitched tones of 5000 Hz (*A*) to 2000 Hz (*B*), but greater at 1000 Hz (*C*). Redrawn, with permission, from Kolz and Johnson (1981).

90–100 per min for monitoring small, fast-moving animals, a rate of 50–70 pulses per minute is usually adequate.

Since power output varies with antenna dimensions, this parameter should only be given for ready-to-mount tags. It is in any case likely to vary between the tags, if only because of differences within any batch of components.

Transmitters manufactured in the United Kingdom, and in some other countries, have to be Type Approved, which indicates that they meet certain government specifications (Skiffins, 1982). For instance, to receive Type Approval from the United Kingdom Department of Trade and Industry, transmitters must not radiate excessively on unwanted harmonics of the frequency set by its crystal. The crystal is a thin slice of quartz, which oscillates on a very precise frequency. This fundamental frequency is inversely proportional to the thickness of the quartz slice. Since the quartz cannot be cut thin enough for a fundamental much above 20 MHz, higher frequencies are obtained by cutting so that the crystal oscillates more on an overtone than on the fundamental frequency. Thus, a crystal for a 150 MHz tag would probably have a fundamental frequency of 16.7 MHz, but the quartz would be cut in such a way that it tended to oscillate on the third overtone: it would be supplied as a 50 MHz crystal. The transmitter circuitry would then "multiply" this frequency by a whole number, in this case 3, to give 150 MHz. However, there would also be a tendency for signal production at other harmonics of 50 MHz: at 50 MHz (first harmonic), 100 MHz (second harmonic), 200 MHz (fourth harmonic) and even higher values.

2. Basic Equipment

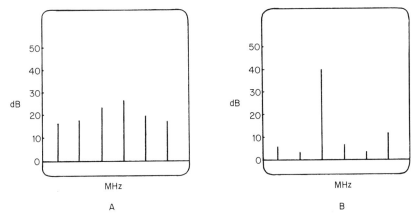

Fig. 2.9. Spectrum analyser output from two radio tags. One tag (*A*) is poorly tuned, radiating strongly on all six harmonics that are visible on the display, and more strongly on the fourth harmonic than on the intended third harmonic. The other tag (*B*) is well tuned, radiating much more strongly on the third harmonic than at any other frequency.

These harmonics can be examined on a spectrum analyser, as shown in Fig. 2.9.

The spectrum on the left of this figure is from a transmitter which was radiating strongly on several harmonics, and produced more signal on the fourth harmonic than on the third, for which it was intended. The transmitter was grossly inefficient, wasting power on unwanted harmonics. In contrast, the transmitter on the right meets the requirements for Type Approval (to MPT 1309 and MPT 1312), by radiating at least 25 dB more on the intended third harmonic than on any other. Without access to a spectrum analyser, there is no easy way to assess the tuning of transmitters which lack Type Approval. Nevertheless, it is worth looking at the inductor coils (Chapter 3) of single-stage tags which are potted in transparent material: tags with neat, circular coils, are likely to be more efficiently tuned than those with loose, splayed windings.

Most manufacturers stipulate the sort of range and life to be expected from a given tag, with a particular receiving system, but there is often considerable variation within each batch of tags. Life may be reduced by poor battery quality, which the manufacturer has been unable to detect, and range may be reduced by your particular terrain or receiving system. For instance, poor connections to the receiving antenna, either in the antenna cable or within the receiver itself (a failed centre-pin connection is a frequent fault in LA-12 receivers), can reduce the range to less than a quarter of the expected value. Do not immediately blame the tag manufacturer if the range is far too little from all your tags! Do, however, return single tags which are sub-standard, and do contact the firm if you have other complaints. If you decide to change supplier on the basis of an initial problem, without seeking an explanation, you may have the same problem with the new firm. In demanding applications, it often pays to work closely with one manufacturer until the problem is solved. In the same way, let the manufacturer know if you are particularly pleased with the product, if only because this may get you a more sympathetic response when you next submit a last-minute order.

V. TAG DESIGN

With purchased or home-made tags, other important considerations are the power source, the antenna format, the mounting technique, and possible sensor options.

A. Power sources: primary cells

Most wildlife tags are driven by primary cells, typically mercuric or silver oxide cells, organic or inorganic lithium cells (reviewed in Ko, 1980), or zinc–air cells. Table I shows the voltages of these cells, and compares the energy densities of some which are used for radio tags in the United Kingdom. There is a list of suppliers in Appendix II.

The high energy density of large lithium-based cells, at 0.5 or more W h ml^{-1} (and >0.25 W h g^{-1}) makes them preferable for tags with a cell capacity greater than about 500 mAh. Cells with a lithium anode and thionyl-chloride anodic material have the highest voltage (3.6 V) and energy densities, while 1.5 V lithium/copper-oxide cells are ideal for large single-stage tags. Most mercury or silver cells, with zinc anodes and mercuric or silver oxide anodes, have energy densities below 500 W h kg^{-1}. Although high capacity mercury cells can be used with 1.35 V transmitters, to save cost when low weight is not at a premium, these should be treated with caution (Keuchle, 1982). The Mallory RM-1, a 1000 mA h mercury cell weighing 12 g, was a satisfactory inexpensive power source for single-stage tags, but the RM-1/PX-1 is not: it is designed for high current drain, and powers the tags for only ⅓–½ as long as the RM-1. Zinc/manganese-dioxide cells, sold over-the-counter as "long-life" batteries, sometimes been used as an alternative to lithium cells where low cost is more important than low weight and volume.

There are plenty of lithium cell types with a capacity below 500 mA h. Neverthless, mercury, silver and zinc–air cells with an equivalent capacity are often preferable, at least for single-stage tags which will not be stored long before use, because they are less bulky than the small lithium cells and therefore keep down the weight of potting. However, lithium cells are best for low-temperature work: the voltage of mercury and silver-oxide cells can drop enough to stop tags working much below 0°C. Silver-oxide cells are best for the very smallest tags. Their output is 1.5 V, which means that they give slightly more powerful signals than the 1.35 V mercury cells in small tags, with less risk of failure due to voltage drop. The smallest silver-oxide cells also have higher energy densities than their mercury equivalents (see Table I).

Table I. The voltage, capacity, dimensions and energy density of some cells suitable for use in radio tags.

Source	Model	Type	Voltage	Capacity	Diam. × height	Weight	W h g^{-1}	W h ml^{-1}
Rayovac	RW310	Zn/AgO	1.5 V	24 mAh	8×2 mm	0.3 g	0.12	0.45
Rayovac	RW311	Zn/AgO	1.5 V	39 mAh	8×2.5 mm	0.5 g	0.12	0.45
Rayovac	RS13G	Zn/AgO	1.5 V	75 mAh	8×5 mm	1.1 g	0.10	0.45
Rayovac	R212	Zn/HgO	1.35 V	16 mAh	5.5×3 mm	0.3 g	0.08	0.27
Rayovac	R312	Zn/HgO	1.35 V	45 mAh	8×3.5 mm	0.7 g	0.08	0.37
Rayovac	R13	Zn/HgO	1.35 V	85 mAh	8×5 mm	1.1 g	0.10	0.45
Rayovac	RP675	Zn/HgO	1.35 V	220 mAh	11.5×5 mm	2.7 g	0.11	0.53
Rayovac	R625	Zn/HgO	1.35 V	350 mAh	15.5×6 mm	4.2 g	0.11	0.41
Rayovac	RP401	Zn/HgO	1.35 V	950 mAh	12×28.5 mm	11.3 g	0.11	0.41
Activair	A312	Zn/O	1.45 V	90 mAh	8×3.5 mm	0.6 g	0.22	0.77
Activair	A13H	Zn/O	1.45 V	170 mAh	8×5 mm	0.9 g	0.27	0.95
Activair	A675H	Zn/O	1.45 V	400 mAh	11.5×5 mm	1.7 g	0.34	1.04
Saft	LC07	Li/CuO	1.5 V	400 mAh	14.5×12 mm	4.5 g	0.13	0.28
Saft	LC02	Li/CuO	1.5 V	1.4 Ah	14.5×25 mm	7.3 g	0.29	0.49
Saft	LC01	Li/CuO	1.5 V	3.3 Ah	14.5×50 mm	17.4 g	0.28	0.62
Sanyo	CR-1/3N	Li/MnO$_2$	3.0 V	160 mAh	11.5×11 mm	3.3 g	0.14	0.42
Sanyo	CR-14250	Li/MnO$_2$	3.0 V	750 mAh	14.5×25 mm	9.0 g	0.25	0.54
Sanyo	CR-17450	Li/MnO$_2$	3.0 V	2.2 Ah	17×45 mm	24 g	0.27	0.65
Sanyo	CR-23500	Li/MnO$_2$	3.0 V	4.5 Ah	23×50 mm	50 g	0.27	0.65
Saft	LS3	Li/SOCl$_2$	3.6 V	850 mAh	14.5×25 mm	8.5 g	0.36	0.72
Saft	LS6	Li/SOCl$_2$	3.6 V	1.8 Ah	14.5×50 mm	15.5 g	0.42	0.81
Saft	LSH14	Li/SOCl$_2$	3.6 V	5 Ah	26×50 mm	56 g	0.32	0.69
Saft	LSH20	Li/SOCl$_2$	3.6 V	10 Ah	33×61 mm	120 g	0.30	0.70

Note that equivalents for most of these cells are available from other manufacturers. For instance, Duracell RM series mercury cells (from FEC) are equivalent to the Rayovac models, while Tadiran lithium/thionyl chloride cells can replace those from Saft.

Zinc–air cells use an ingenious catalyst system to combine anodic zinc with atmospheric oxygen, generating 1.5 V. Since the cathode is replaced by a thin chamber with one or more airholes, there is much more room for the zinc anode than in the mercury or silver cells, and the zinc–air cells have about twice their energy density. Zinc-air cells have been quite successful on bats and on non-aquatic birds, using the high-current versions of Activair sizes 312, 13 and 675, but have proved unsuitable for tags on small rodents and reptiles, presumably because the airholes get blocked. Commercial tags with these cells are simple to start in the field, by removing an adhesive strip which covers the cell's airhole. This avoids the inconvenience of soldering, or the cost and extra bulk of magnet-operated reed-switches, which can be used to start tags with other types of cell. Once opened, however, zinc–air cells lose capacity even if no current is drawn, because their electrolyte dries out. Gould-Activair give an operating life of two months for opened A675H cells. This is a conservative figure, intended for cells which are used in relatively dry air: these cells run one-stage tags on birds for 12–20 weeks in Britain. Maximum lives are obtained if one of the two A675H airholes is left sealed, to reduce fluid loss from the electrolyte, and one such cell ran a very low current tag for nearly nine months. For comparison, the Duracell MP675 mercury cell runs low current tags for only 10–14 weeks, but is more reliable in the field than the zinc–air types.

It is important to realize that all these small cells have rather short shelf-lives. They tend to leak electrolyte at the edges of the narrow insulating strip between anode and cathode, especially if exposed to large temperature fluctuations. Mercuric oxide cells have sodium or potassium hydroxide electrolyte, which forms crystals of bicarbonate with carbon dioxide and water in the air. Although these crystals are non-conducting, it is wise to discard cells which show such signs of leakage. In general, it is best to store these small cells at low, stable temperatures (e.g. in a refrigerator), and not to keep them for more than a year. Tags which contain these cells should not be stored for more than a few months: any leaking electrolyte will remain conducting, in the absence of air, and not form tell-tale crystals. Cells with an anode–cathode gap of at least 1 mm store better, for several years with little capacity loss in the case of lithium types in low humidity.

On the whole, dividing the specified cell capacity by the transmitter current tends to overestimate tag life, but not too seriously for the cells in Table I. If you wish to use other types, you should check with the manufacturers that they are suitable for low-drain pulsed operation.

B. Solar power

As an alternative to primary cells, photoelectric panels can power wildlife tags for several years (Patton *et al.*, 1973; Church, 1980). The smallest tags, weighing only 6 g (single-stage) to 15 g (two-stage) are powered by photoelectric panels alone, and therefore stop transmitting in low light intensities. Such tags are not very satisfactory for systematic studies, because loss of signal may mean that the transmitter is heavily shaded rather than out of range, and the signal also fails at night, in nest-holes, or if the animal lies dead on its back. Slightly heavier packages can be built with 20 mA h nickel-cadmium cells, which are charged by the solar panels to provide power in poor light. However, NiCads subject to frequent unregulated charging tend to malfunction within a couple of years, after which the tag depends solely on the solar panels (Ko, 1980; Keuchle, 1982).

All this means that solar-powered tags are much more suitable for some projects than for others. For a start, they are best attached dorsally, to diurnal species which live in fairly open habitats. Single-stage solar-powered tags are at present most appropriate for animals which weigh at least 100 g (at which weight they can carry a 6 g tag) and not more than 250 g (at which weight they can carry a 12 g tag with a lithium cell giving a life of nearly two years). Considering the risks of antenna breakage through corrosion or metal fatigue, of moisture penetration, and of the animal dying, it is seldom reasonable to expect a life of more than two years from radio tags on such small animals.

The best use of solar-powered radio tags is for animals that require two-stage or even more powerful transmitters. Even a low-power (150 μA) two-stage circuit requires about 25 g of lithium cell (a 35–50 g tag) for a two-year life, whereas two-stage solar-powered tags with rechargeable NiCads can weigh as little as 17 g (Church, 1980). Such transmitters have therefore been used mainly as backpacks on birds weighing more than 350 g, especially those which live in open country and have relatively large ranges. They are also invaluable as long-life patagial tags for very large birds (Ogden, 1985), and for ear-tagging young mammals which would outgrow radio collars (Serveen *et al.*, 1981). Solar power can also be used to power 160 g tags which are tracked by satellite (Fuller *et al.* in press; see section VI).

Future wildlife tags may well use solar power with back-up lithium cells, or with circuitry which regulates the charging of NiCads and thus increases their reliability.

C. Antenna format

The main types of antenna used for animal tags are loaded loops, tuned loops and whips. Whips are the simplest antennas to construct. Nickel guitar wire, or nylon-coated multistrand stainless steel fishing trace, is commonly used for small tags, with multistrand steel cabling covered by tough heat-shrink tubing for larger animals. Systems with a main whip and a ground-plane antenna are the most efficient antennas commonly used on animal tags, provided that whip length can be at least $\lambda/8$. The ground-plane antenna is typically a second whip, ⅔ the length of the main whip antenna, and perpendicular to it or in the opposite direction (Plate 4).

If tags must be compact, however, for implants or on animals which would damage protruding whips, loop antennas are often preferable. Loaded loop antennas, made of iron-dust or ferrite cores wound with several turns of enamelled copper wire, have been used quite extensively for transmitters in the 104 MHz band and below (Taylor and Lloyd, 1978; Reeve, 1980) because of the stability they confer. Since they are small enough to be completely embedded in the tag, they are less prone to detuning than exposed antennas. The detuning of exposed antennas, through capacitive effects due to proximity and movements of the animal, seems to stop transmitters relatively easily at 104 MHz, compared with the circuits described in Chapter 3 for 140–220 MHz.

Simple tuned-loop antennas are the most practical compact alternative to whips for the higher frequency band. A loop of wire or brass strip is often used to form a collar on small to medium-sized mammals, the upper size limit being determined by the maximum size easily tuned to resonance at the

Plate 4. A 10-month tag on two goshawk tail feathers. Note the free-standing ground-plane antenna at the front of the tag, and the main antenna which is being tied and glued along the shaft of one tail feather.

frequency concerned. This limit is a diameter of about 5.5 cm for a 10 mm wide strip of 1 mm thick brass at 173 MHz, being greater at lower frequencies and vice versa. Using thinner collar materials raises the limit somewhat, but also reduces the antenna efficiency at a given diameter. For implants, or other tags with embedded antennas, the most common format has been a single- or double-loop antenna, around the whole tag or at one end. However, the larger tags can also be built with a spiral whip antenna just below the surface of the potting.

D. Mounting techniques

In choosing tags designed for a particular attachment technique, one needs to be aware of the associated advantages and limitations. The mounting technique affects the size of tag which can be attached, its radiation efficiency, its influence on the animal and the skill needed to attach it. These aspects of project planning are discussed now, leaving the details of tag construction and attachment until Chapters 4 and 5.

Tags are attached to birds mainly with harnesses, tail-mounts, leg-mounts and ponchos or necklaces, all of which have whip antennas. Harness-mounted backpacks are the only feasible tags for many projects. They are placed close to the bird's centre of lift, and may therefore be heavier than the alternative types. Nevertheless, harness-mounted tags have been shown to have an adverse effect on plumage, behaviour and even survival in a wide range of species (Nicholls and Warner, 1968; Boag, 1972; Ramakka, 1972; Greenwood and Sargeant, 1973; Lance and Watson, 1977; Johnson and Berner, 1980; Perry, 1981; Small and Rusch, 1985). Work on gulls, using dummy tags of various weights, suggested that the adverse effects stem mainly from the harness (or handling process) rather than the tag itself (Amlaner et al., 1978), although this study also recorded a decrease in brood survival of parents with higher tag weights (see also Chapter 5). Adhesive mounting of backpacks seems to be a safer alternative (Graber and Wunderle, 1966; Raim, 1978; Perry et al., 1981), and has been used to attach tags reliably for several months to species whose feathers do not pull out easily (R. Green, personal communication). Although other birds (especially passerines) have shed such tags in a few days, rapid tag detachment can be an advantage if many nests or roosts are to be found by re-using a small number of tags. Tags have also been sutured to bird backs (Martin and Bider, 1978), but such an invasive technique seems best avoided.

Tail-mounts are preferable to backpacks, provided that the tail feathers are large and firm enough, and the tags need not outlast the moult (Bray and Corner, 1972; Dunstan, 1973; Fuller and Tester, 1973; Fitzner and Fitzner,

1977; Kenward, 1978). They should not weigh more than about 3% of bodyweight. Compared with backpacks, however, weight is saved through lack of a harness, and by using lightweight antenna wire if this is secured by binding it along the shaft of a rectrix (Plate 4). Like glue-on backpacks, tail-mounts also detach safely, through moulting at the latest, so that tags which are still transmitting can be recovered and re-used.

Leg-mounts have been used on large, long-legged birds, such as storks (A. Burnhauser, personal communication) and cranes (Melvin *et al.*, 1983). They are also useful for tagging nestlings (Plate 5), whose growing feathers might be damaged by harnesses or tail-mounts (Kenward, 1985). However, leg-mounts often have less range than the other types, because they are closer to the ground and have shorter antennas.

Ponchos, which evolved from a visual marker worn round the neck, are a good long-term tag for some birds (Amstrup, 1980). The movements and

Plate 5. A goshawk nestling tagged on the leg. If backpacks are put on hawks at this age, the young birds probably habituate to them before fledging. (Photo by M. Karlbom.)

survival of ruffed grouse with ponchos were better than for grouse with harness-mounted tags (Small and Rusch, 1985). Similarly, a necklace design has proved much simpler than harnesses or tail-mounts to attach to pheasants (Fig. 2.7), having less adverse affect on the birds than the former, and less tendency to detach prematurely than the latter. The necklaces also give the best detection ranges, partly because their main antennas are vertical and partly because of a ground-plane antenna in the necklace cord. Ponchos or necklaces seem ideal for game-birds. However, the main antenna can get tucked under the wing and thus irritate birds which spend much time in flight, and these designs would be unsafe for birds which swallow food items nearly as large as their heads. They should also probably be avoided for waterbirds at high latitudes, where neckband visual markers can accumulate lethal masses of ice (Zicus *et al.*, 1983).

Other tag-mounting techniques for birds include the marking of diving species with nasal saddles (Swanson and Keuchle, 1976) or implants (Boyd and Sladen, 1971; Korschgen *et al.*, 1984), which avoid the drag and possible insulation loss from harnesses, and the use of solar-powered patagial tags on condors (Ogden, 1985). Patagial radio tags are probably only suitable for species which flap their wings slowly and relatively little. Finally, transmitters have been built into dummy eggs, either to record incubation parameters (Schwartz *et al.*, 1977) or to help identify nest predators.

The typical mammal tag is a collar (Plate 6). For large mammals, the tag itself is usually bound, bolted or riveted to a collar of leather, braided nylon strapping, or other flexible plastic. If grooming is likely to damage an

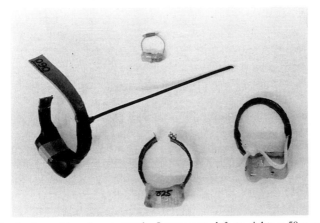

Plate 6. A selection of tags for mammals. Lower row, left to right: a 50 g collar with a magnet-operated switch, for a marten (*Martes martes*); a 15 g open tuned-loop brass collar for a red squirrel (*Sciurus vulgaris*); a 20 g closed tuned-loop brass collar, with cable-tie fastening, for a grey squirrel (*Sciurus carolinensis*). Top: a 2 g "sliding-8" tuned loop collar for a small rodent.

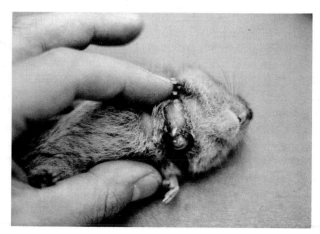

Plate 7. A bank-vole (*Arvicola terrestris*) with a "sliding-8" collar.

exposed whip antenna, a whip may be sandwiched between two strips of collar material, or an antenna embedded within the tag. Smaller mammals, which would not damage or be impeded by exposed whip antennas, may be equipped with necklace tags. The low weight and soft necklace cord makes these tags ideal for easily-shocked species like hares. Tags can be built with a zig-zag wire whip between two strips of leather, or with a short length protruding at the back of the neck, or with braided copper sleeve which is held round a cable tie by heat-shrink tubing. For small mammals, however, the most efficient alternative to an exposed whip is likely to be a tuned-loop collar.

For the smallest mammals the tuned-loop collar is usually of PTFE-coated wire, which is tightened round the neck with a metal crimp or heat-shrink tubing (Plate 7). Animals above about 150 g can carry collars of fine brass strip or multi-strand brass wire (picture wire), which give stronger signals than a fine wire collar with a similar circumference. The most efficient tuned-loop collars have a fixed circumference. This may be a continuous metal loop, which is slipped over the animal's head before inserting a stop or tightening a cable-tie (Plate 8) to reduce the internal circumference (Wood, 1976). Fixed-circumference collars which bolt or crimp together round the neck are slightly smaller, and therefore a little less efficient, but also weigh less and hang down less from the animal, thus being more suitable than continuous-loop types for animals which creep close to the ground. Moreover, they start transmitting only when the loop is closed, so that no soldering or switching is needed to start them. The same applies to brass-strip collars which contain several bolt holes, so that their circumference can be varied to suit a range of head and neck sizes. The tuning of these variable circumference collars is less precise than for fixed-circumference types, unless a variable capacitor can be "tweaked" once they are on the animal — no easy task, unless the mammal is anaesthetized.

The use of collar tags is difficult or impossible on mammals with very short or tapering necks, especially mustelids and species streamlined for swimming or burrowing. On mink, for instance, supple leather collars with sandwiched whip antennas seem less likely to be shed than brass tuned loops, which are more efficient but too rigid to give a really close fit (N. Dunstone, personal communication). Moreover, it is best not to use collars with inflexible brass strips on mammals which live in rocky holes. Rapid loss of signals from such tags on wild martens (fishers), despite satisfactory operation on animals in enclosures (Skirnisson and Feddersen, 1985), suggests that the tags were either damaged or fatally wedged in the dens.

Since collars will not stay on otters or many seals at all, tags for these animals have been attached to harnesses or glued to the fur using epoxy resins (Broekhuizen *et al.*, 1980; H. Kruuk, personal communication; K. Nicholson, personal communication). It is difficult to build harnesses which

Plate 8. Mounting a tag on a grey squirrel.

will not eventually abrade these animals, and adhesive mounting cannot be relied on to last more than two or three months, at best. In such cases, intraperitoneal implants appear to be the best solution for long-term tagging (Melquist and Hornocker, 1979; Eagle *et al.*, 1984; Davis *et al.*, 1984). Marine mammals have also had tags attached to their fins or flippers (Butler and Jennings, 1980; Garshelis and Siniff, 1983), or to hooks or miniature harpoons in their blubber (Watkins *et al.*, 1981; Mate *et al.*, 1983). The tags should be attached where they are most exposed when the animal surfaces.

Adhesive mounting has become the preferred technique for tagging small bats (Williams and Williams, 1967; Stebbings, 1982), using epoxy resin or silicone rubber to glue tags with whip antennas to the back, close to the centre of lift. A necklace provides a durable attachment for the larger bats, and can be equipped with a ground-plane antenna to increase the signal strength.

Some mammals have horns or other protuberances to which tags can be attached. Pasty silicone sealants give flexibility for attaching transmitters to hedgehog spines (Morris, 1980), while Anderson and Hitchins (1971) embedded tags in rhinoceros horns and Pages (1975) wired them to pangolin scutes. Tags have also been fixed to mammal ears (Serveen *et al.*, 1981), and tail-bases (M. Gorman, personal communication).

Many large reptiles have horny scutes which can be drilled to anchor attachment wires. Tags have been bound and glued in this way to turtle and tortoise carapaces (Schubauer, 1981; I. Swingland, personal communication), and to the heads or backs of crocodiles (Smith, 1974; Yerbury, 1980). Small reptiles are less easy to tag, although harnesses have been used on lizards (Fullager, 1967), and adhesive mounting, with surgical tape and cyanoacrylate glue, for the temporary attachment of tags to snakes and small lizards (A. Cooke and A. Gent, personal communication). For long-term radio monitoring of snakes, there is no real alternative to surgical implantation (Barbour *et al.*, 1969; Madsen, 1984). The same often applies to amphibians (Stouffer *et al.*, 1983; Smits, 1984) whose skin is not only too damp for adhesive mounting but is also rather easily damaged by harnesses. However, frogs and toads and snakes have been persuaded to swallow transmitters in some projects (Osgood, 1970, Fitch and Shirer, 1971; Brown and Parker, 1976). Ingestion of a large object might be expected to influence subsequent behaviour, but toads containing relatively large tags continued feeding normally in at least one study (M. Swann, personal communication).

Four main techniques have been used for radio-tagging fish (Winter *et al.*, 1978). Tags have been attached externally to the dorsal surface, either beside the dorsal fin or mid-dorsally on a saddle mount behind this fin, by sewing absorbable sutures (Solomon & Storeton-West, 1983) or "ground-plane" wires through the dorsal musculature. Such tags are streamlined and have short whip antennas, which are relatively more efficient in water than in air (Weeks *et al.*, 1977). Tags can also be inserted into the stomach, where they have been retained for several weeks by salmon in fresh water (Solomon and Storeton-West, 1983). Careful insertion is required to avoid damage to the stomach, and the risk of rupture increases as the stomach lining atrophies while salmon move upstream to spawn (Hayes, 1978). Since external attachments can abrade the skin and increase the risk of fungal infection, the most satisfactory technique for long-term tagging of most species appears to be implanting in the abdominal cavity. Implants can have whip antennas threaded with a cannula under the skin (Winter *et al.*, 1978), in which case the reception ranges may be only 10% less than for external tags, provided that they are properly tuned.

Implanting thus appears to be the most humane and effective technique not only for tagging fish and amphibia, both of which have easily-damaged skin, but also for some other animals which risk entanglement, lethal heat loss or starvation if equipped with harnesses.

E. Sensors

Tags for movement or temperature telemetry are straightforward to construct (Chapter 4), and are therefore widely available commercially. They use a mercury switch or a thermistor, respectively, to alter the pulse rate (usually making the pulses shorter as well as more frequent).

Mercury tilt-switches, containing a drop of mercury which rolls to bridge internal contacts, are used to indicate movement, posture and mortality. In the simplest tags, a low-value pulse capacitor gives a short and frequent pulse when the switch is open. Closure of the switch puts a second capacitor in parallel with the first, thereby increasing the pulse duration and decreasing the pulse rate. Such switches, containing mercury which rolls longitudinally, have been set at about 45° to the horizontal in transmitters on goshawk tails (Fig. 2.10). When a hawk is perched, with its tail close to the vertical, the mercury switch is closed and a 30 ms pulse occurs about once per second, a convenient pulse duration and rate for radio location while the hawk is in its most common posture. When a hawk flies, the tail is horizontal and the switch open, producing 15–20 ms pulses at about twice the previous rate. The flight signal also varies in amplitude, as the hawk's direction and hence the antenna orientation changes or as obstacles come between hawk and receiver, so that this signal can be distinguished by ear from the steady fast pulse given by an incubating hawk or one lying dead. Feeding produces a characteristic signal too, alternating between a slow pulse as the hawk arches

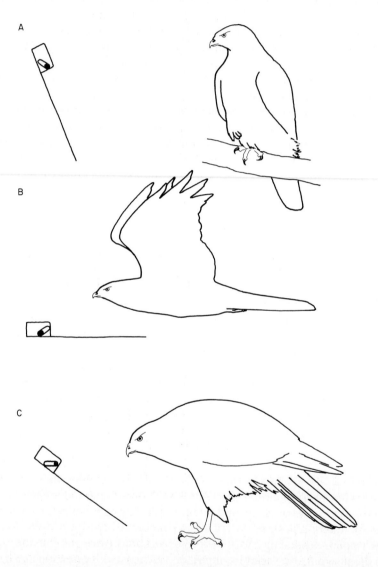

Fig. 2.10. Mercury switches are used as posture sensors in tags tail-mounted on hawks. The tail points downward on resting hawks, and the switch is set for a slow signal pulse (*A*). In flight the mercury moves to its other position, and the signal pulses are fast (*B*). When a hawk pulls at food, the the tail is raised and lowered, thus slopping the mercury from one position to the other and giving an irregular signal (*C*).

its back to pull at food, and a fast pulse when the tail becomes more horizontal as the bird relaxes to swallow (Kenward *et al.*, 1982).

On many species, tags cannot be mounted firmly enough or in suitable places for sophisticated posture monitoring, but can be used to detect movement. Thus a mercury switch may be mounted transversely at 5° to the horizontal in the tag on a tortoise or hedgehog, giving a steady slow signal when the animal rests, but an irregular pulse as its rolling gait slops the mercury, thus opening and closing the switch. On fast-moving mammals the switch may be set fore-and-aft in a collar, at 5° to the horizontal and closed when the collar hangs level. The slight accelerations and decelerations when the animal moves are enough to make the signal irregular. Some firms make tags which are sensitive to movement in any direction: they contain tipover switches, where the mercury can roll in any direction from central electrodes. These devices are less suitable for posture sensing than switches with a longitudinal action.

In motion-sensing tags, a steady signal when the animal should be active may mean that it is dead, and this feature is exploited in special mortality-sensing tags. These contain an integrated circuit, which changes the pulse rate after a preset delay in which the mercury switch has not moved (Kolz, 1975; Keuchle, 1982). A delay of hours rather than minutes should be chosen, or there will be many false alarms as the animal takes short rests. Such tags have also been used for activity recording, but tags with simple tilt-switch circuits tend to be more satisfactory (Gillingham and Bunnell 1985). However, combined activity/mortality tags could be built, with three types of pulse: two during movement and a third which occurs only after prolonged lack of movement. It will soon be possible to register activity and mortality automatically without timer chips in tags, by using inexpensive "intelligent" loggers, based on portable microcomputers, to detect when the received pulse rate of simple mercury-switch tags has not changed during a programmed delay.

Temperature-sensing tags are based on thermistors. These are semiconductors which act as heat-sensitive resistors, typically with a positive temperature coefficient: the resistance decreases as the temperature rises. In the simplest tags, this drop in resistance allows a greater charging current to flow into the pulse capacitor, thus increasing the pulse rate as the temperature rises. Since the thermistor replaces one resistor in the oscillator circuit, such a transmitter need weigh no more than in an unmodulated tag. However, because the sensor changes the current through the whole circuit, there is a trade-off between the temperature sensitivity, the range of temperatures measurable, and the cell-life. Sufficient current must still flow for the transmitter to pulse when thermistor resistance is high, at the lowest likely temperature, and this will be similar to the current through a low-power unmodulated tag. Assume that the thermistor tag has a 2.5% change in pulse-rate per C°, enough for 1–2°C accuracy in field temperature measurements, and operates over a 40°C range. If the thermistor has a more or less linear temperature response in that range, the circuit current and pulse rate at 40°C will be double that at 0°C (since 40 × 2.5% is a 100% increase). If the tag operates mainly at the higher temperature, for instance in a mammal which only occasionally enters torpor, then the cell life will be about half that of the unmodulated tag. Similarly, doubling the sensitivity of such a tag would halve the cell-life again, unless one decided to use only the upper half of the temperature range. The pulse-rate would then change 5% per °C, but the tag would cease to pulse below 20°C.

A thermistor can also be used to leak current across one of two serially connected capacitors, which replace the pulse capacitor. As in tags with mercury switches, this provides pulse modulation without affecting total circuit current. Although the capacitive subcircuit can be built into the smallest tags, to provide temperature sensing with no cell-life reduction and perhaps 50 mg weight increase, experimental tags have proved less sensitive than the previous type and more susceptible to de-tuning: the pulse rates were affected as much by changes in antenna position as by small temperature changes.

A third option is to link the thermistor to a pulse-generating circuit (e.g. an astable multivibrator) whose current is essentially independent of the transmitter's output circuitry (e.g. Amlaner, 1978; Anderka, 1980). Such tags provide the most accurate temperature measurement, and are more or less essential for external tags which have long leads to implanted thermistors. However, the additional circuitry adds weight, draws some current and increases construction costs.

An obvious use for thermistor tags is to measure body temperatures, using implanted tags, ingested tags or external tags with implanted sensors. Another use is to measure ambient temperature, perhaps to see what microclimate an animal prefers, but also to monitor behaviour. For instance, thermistors attached to harness straps under a bird's wing are cooled when the bird takes flight, from close to body temperature to ambient air temperature (Kenward *et al.*, 1982). This is useful for monitoring the activity of species whose tails and bodies have the same tilt in flight and at rest. Similarly, thermistor collars on small mammals can indicate whether they are curled up in warm nests or out foraging (Osgood, 1980; Kenward, 1982a), and thermistor tags on all warm-blooded animals also indicate mortality (Stoddart, 1970). In these applications, temperature sensitivity can be reduced to prolong cell-life.

There are a number of other sensors which have been used less widely in wildlife radio tags. Changes in resistance or capacitance caused by water or urine can reveal aquatic or marking behaviour (Charles-Dominique, 1977).

Ambient light intensity can be monitored with photovoltaic panels, and proximity to other objects or animals can be detected if they are equipped with magnets to operate a reed-swich on the radio-tag. Sophisticated circuitry is needed in tags whose pulse rate is to be changed by pressure sensors (Keuchle, 1982), strain gauges (Fuller *et al.*, 1977) or compasses (in acoustic tags: Mitson *et al.*, 1982), and this is also true for tags triggered by heart beats (Gessamen, 1974, Wolcott, 1980b). It is difficult to ensure that the latter tags are triggered only by ventricular contractions, and not by the muscle potentials resulting from auricular contractions and trunk musculature as well. With long leads from implanted electrodes to external transmitters, it is hard to maintain accurate pulse triggering for more than a few days.

Complex circuits are also required for tags which can telemeter information from more than one sensor, for instance by modulating pulse duration and pulse interval, or by sending a train of pulses in which each is modulated by a different sensor (Smith, 1974; Standora, 1977; Lotimer, 1980). Since it is no simple matter to assemble receiving equipment which can interpret these signals, the use of multiplex tags has so far remained the province of groups working with electronics engineers. However, suitable wildlife tags and receiving packages will probably become available commercially in due course. This could also apply to frequency-modulated tags which contain microphones, to transmit calls or feeding noises (Greager *et al.*, 1979; Gautier, 1980).

F. Data storage

Work on marine mammals and some diving seabirds, from which radio tags can transmit a signal very infrequently, has been a strong incentive for developing tags which can store information on dive depths, durations and physiology for transmission when the animal surfaces. A crude tag of this type, containing sensors, analogue-to-digital conversion and logic circuitry, clock and memory, costs about £300 to build (Robinson, 1986). Present designs transmit relatively small quantities of data in repeated signal streams, which are triggered by a pressure sensor as the animal surfaces or by a master transmission (i.e. in a transponding mode). Future designs could store large quantities of behavioural data, perhaps dumping these once a day to automatic receiving stations at bird or bat roost sites or near carnivore dens.

G. On-off options

Many home-built long-life tags are transmitting when they are potted up, shortly before use. Others are simple to start in the field, for instance by joining a tuned collar loop, by removing the tape to let oxygen into a zinc–air cell, or by exposing solar-powered tags to the light. However, there are at least four more options for switching on tags.

When funds and cell-life are at a premium, the transmitter circuit is usually completed by soldering two exposed wires. The join should then be coated with epoxy resin adhesive, or other suitable sealant, to prevent moisture penetrating the package along the wires. A portable soldering iron is needed to start short-life tags in the field: the Oryx butane-powered iron (from RS or Biotrack; see Appendices I and II for addresses) is lighter and smaller than battery-powered types. If the wires are merely twisted together, their junction resistance may eventually increase enough to stop the transmitter.

Tags can also be built with a little protruding wire loop, which makes a connection to shut down the oscillator circuit transistor (Macdonald and Amlaner, 1980). Such a tag transmits when the wire loop is severed. It is simple both to test the tag, by cutting and then rejoining the loop, and to start the tag in the field, by removing the loop entirely. The drawback is that a transistor leakage current continues to flow through the circuit when it is shut down, at 2–4% of the operating current, so this option is only suitable for short-life tags if they are to be used fairly promptly. It is worth noting that a similar leakage current flows through tuned loop tags which start by completing the loop.

For a 1–2 g weight penalty and some extra expense, tags can be equipped with a magnet-operated reed switch. The normally-closed (NC) switch remains open while a magnet is in a marked position on the tag's outer surface, and no current flows in the transmitter circuit until the magnet is removed. This option is especially suitable for very long life tags, which may need to be stopped and restarted several times before their cells are exhausted, or for tags which are very thoroughly encapsulated for use on large or aquatic animals.

The fourth option is to use tags with CMOS latching circuitry, which can be switched on and off to conserve cell life by photosensors (at night), by temperature sensors (during torpor), by fixed-interval timers (Smith, 1980), or by seawater between exposed metal contacts (Broekhuizen *et al.*, 1980). The disadvantage with these options is the extra tag cost, and the risk of signal loss if, for example, animals with photosensors enter holes or die with the tag covered.

VI. TAGS FOR TRACKING BY SATELLITE

Tags for tracking by satellite have been used since the early 1970s, on wapiti (Buechner *et al.*, 1971), polar bears (Kolz *et al.*, 1980; Schweinsburg and

Lee, 1982), turtles (Timko and Kolz, 1982) and basking sharks (Priede, 1980). The early tags weighed 5–11 kg, for location by the Nimbus 3 and Nimbus 6 weather satellites, and were therefore suitable only for large animals.

Satellite tracking works on the Doppler principle. A frequency shift in each received signal indicates the satellite's speed relative to the tag, and the tag's direction is computed from the ratio of this speed to the satellite's true ground speed. For example, a ground-speed/relative-speed ratio of 1:2 would give a tag bearing of 45° to the satellite's track. To estimate a fix, at least two uplinks are needed during each pass, which may take as little as 10 min (horizon to horizon). Since the tag's position could then be on either side of the satellite's track, it can only be located unambiguously (i) if it can be recorded again from the different track on another orbit, or (ii) with reference to a recent previous fix, or (iii) if one of the two computed positions is impossible (e.g. for a whale on dry land!).

Since 1978 the Argos system has been made available for animal tracking. This equipment, designed and operated by CNES in Toulouse, is carried on two Tiros satellites of the U.S. National Oceanic and Atmospheric Administration. Argos charges only $10 per day per tag, and tag power requirements are reasonable: a 360–920 ms pulse train, of which the first 160 ms must be constant carrier (at 401.65 MHz) for the Doppler vectoring, at 1–2 W. However, the intended location accuracy of within 5 km requires a very high frequency stability. Signals are rejected if they shift more than 2 Hz during a satellite pass, or 24 Hz between orbits (Priede, 1986). Compare this with the drift of perhaps 50 Hz per °C in normal tracking tags. Moreover, the Argos system requires at least four uplinks over an interval of at least 7 min during a satellite pass.

Tracking by satellite is the most economic technique for wide-ranging marine mammals, although the uplink requirements mean that few fixes are obtained for species which are rarely on the surface. This is because the number of satellite passes per day ranges from only seven, at the equator, to 15 at high latitudes. Satellite tracking also has considerable potential for monitoring migration routes and important feeding areas of large raptors, cranes and seabirds. This potential is likely to be realized, thanks to the recent development of suitable 160 g solar-powered tags, the first of which have been used in the satellite tracking of bald eagles and giant petrels (Fuller et al., 1984).

VII. TAGS AS CAPTURE AIDS

Radio tags can save a great deal of time spent in checking isolated traps (see e.g. Hayes, 1982; Nolan et al., 1984). A simple approach is to use a magnet-operated transmitter: a cord from the trap's door or other other moving part pulls the magnet off the transmitter when the trap is activated. For fail-safe operation the radio beacon should have a normally-open (NO) reed switch, to turn the signal off when the magnet is pulled away.

Tags which fit within anaesthetic darts (Lovett and Hill, 1977), or around the needle of darts without a suitable compartment, can be obtained commercially from some firms listed in Appendix I. Such tags, with ranges of 150–500 m, are useful not only for finding animals which travel some distance before succumbing to the drug, but also for finding darts, containing dangerous drugs like Immobilon, which miss the target.

Radio tags can also be used to locate dens and nests, for instance by laying out tagged carcasses which were scavenged to fox dens (Voight and Broadfoot, 1983), or by dropping beacon tags to help ground-crews during aircraft searches for raptor nests. Once radio-tagged, animals can sometimes be recaught at roost sites, or with drugged food (Huempfner *et al.*, 1975; Kenward, 1976). The ultimate development in this line is a large-mammal collar containing a pair of anaesthetic syringes, with small charges which can be activated by a coded radio signal. These collars have been used to recapture wolves, deer and bears (Mech *et al.*, 1984). They cost about $1200 each, with $350 for the activating transmitter, from Wildlife Materials Inc.

3
Building Transmitters

This chapter describes how to make simple single-stage and two-stage transmitters (see Plate 3). Building transmitters is not something to be undertaken lightly, because a proper set of tools costs around £100, and to ensure really good tuning you also need a spectrum analyser costing £1000 or more. Projects in the United Kingdom require a Testing and Development licence (from the Radio Regulatory Department, Department of Trade and Industry, Waterloo Bridge House, London SE1 8UA). The instructions are intended for biologists with some electronics expertise, and for technicians. Those who intend to make their tags by buying the basic transmitters, and then adding power cells, antennas and mounting materials, should find the information they need in Chapter 4, but would be well advised to read the first two sections of this chapter as well. Those who are interested only in making two-stage transmitters, should read through the construction of the single-stage ones too, because some of the tips given there are not repeated later in the chapter.

There are many different simple crystal-controlled transmitter circuits which are suitable for animal tags, and some authors (marked *) give quite comprehensive construction details (Cochran, 1967*; Kolz and Corner, 1975, Zimmerman *et al.*, 1975*; Taylor and Lloyd, 1978*; Winter *et al.*, 1978*; Cederlund *et al.*, 1979; Macdonald and Amlaner, 1980; Morris, 1980; Thomas, 1980*; Keuchle, 1982; Kolz *et al.*, 1984). The circuits here have been chosen because they require relatively few components, are simple to put together, and tend to give good results straight away if assembled carefully. The individual components are all relatively easy to obtain at the time of writing, although small discrete transistors in particular are unfortunately likely to become obsolete as their main users go over to integrated circuitry. However, the very small size of some components means that they must be ordered direct from specialist manufacturers, rather than the large mail-order distributors. There may therefore be sizeable minimum order charges, or price scaling such that the unit cost for 1–10 components is much more than if 100 or 1000 were ordered. One way round this is to use larger components which are more easily obtained, but this may double or triple the weight of the transmitters. Addresses for component suppliers are given in Appendix II.

The emphasis here is on ease of assembly, and not on making the transmitters as small as possible with the recommended components. By changing the component layout to make the single-stage transmitters as compact as possible, and dispensing with the two-stage printed circuit board (PCB), further reductions in size and weight are possible. This requires considerable skill, however, which can only be acquired over several months through building many tens of tags.

I. TOOLS

You will need a soldering iron which can reach a temperature of at least 420°C, preferably with a needle-tipped bit or one not more than 1 mm across the tip. An RS Components system based on parts 591–297, 544–594, 544–623 and 544–572 is very suitable. This should be used with fine gauge (18–22 s.w.g.) multicore solder.

Precision wire cutters and snipe-nosed pliers can also be obtained from RS Components, as parts 544–370 and 544–409. Artery clamps with the finest possible jaws are ideal for heat-shunting small components, and can most cheaply be obtained from medical suppliers. A ⅛ in or 3 mm drill bit makes a good former on which to wind the inductor coils of single-stage transmitters. Combination pliers (e.g. RS 609–411) are used in a "third-hand" system and also to cut thick wires: the precision wire cutters should be used only for fine component leads, to preserve their edges. A small file is useful for cleaning the surfaces of crystal cans prior to soldering, and a screwdriver with a 2 mm blade is handy for covering the finished transmitter with Rapid Araldite potting.

A small machine vice can be used to hold components which need drilling during tag assembly, and also to make a very convenient third-hand system. Clamp one handle of the combination pliers vertically in the vice, so that they can be opened by pulling the other handle away from the side of the vice, and wind a thick rubber band several times round the jaws of the pliers near their fulcrum. They will then grip, and heat-shunt, crystal cans and the leads of other components. For very fine work, an artery clamp can be gripped in the pliers or vice jaws. These transmitters can be assembled quite easily without bench magnifiers, although some people prefer these aids. Good bench lighting, such as an Anglepoise lamp, is essential.

Minimal test equipment is provided by a (non-digital) multimeter with one of its ranges covering 0–500 μA, and a radio-tag receiver for listening to transmitter signals during tuning. Use the receiver without any antenna, or perhaps with an unwound paperclip in the centre of the antenna socket. If you attach leads with quick-release clips to the meter, and to cells in series to power the transmitters, testing is much quicker than if you have to solder leads each time. Keep the test leads short to minimize detuning effects, and test the transmitters at or below the cell voltage you will use for them. For instance, transmitters for use with nominally 1.5 V lithium/copper-oxide cells should be tested with 1.5 V zinc or alkaline manganese cells, or with 1.35 V mercury cells to be even safer, and not with fresh lithium/copper-oxide cells, which give more than 2 V. Otherwise the tags may stop working within a day or two of being switched on, as the cell voltage falls to its normal operating level. To test that transmitters with 1.35 V mercury cells will run until their power supply is exhausted, you can use 1.2 V nickel–cadmium (NiCad) rechargeable cells.

If you buy untested components for making large quantities of transmitters, you should also make up test circuits with push-in sockets for crystals and transistors. Crystals should be tested at a current slightly below their desired drive level, to ensure that their equivalent series resistance (e.s.r.) is not too high for reliable oscillation. When ordering crystals, you should explain what they are to be used for and preferably send a circuit diagram. Specify that they are for series operation, request a maximum e.s.r. of 35 Ω at an 0.01 mW drive level, and give the required frequency to four decimal places (MHz). Mention also the tag frequency, to the nearest MHz, for which they are intended. A temperature stability of 20–30 ppm to $-10°C$ is generally adequate, with 20 ppm calibration accuracy. Delivery delays, prices and the proportion of satisfactory crystals vary considerably between suppliers.

Transistors too are often very variable in quality. A test circuit which lacks a crystal can be used to check the steady current through each, so that transistors with very poor gain are not used. Those with the best gain will drive crystals which have a relatively high e.s.r. However, they use a relatively high circuit current and thus reduce tag life. If your crystal e.s.r. is low enough, you may therefore prefer to maintain tag life by increasing the circuit resistance when transistor gain is high.

Some 2–3% of cheap mercury switches are defective, too, so it is wise to test these before they are too deeply embedded in potting materials.

II. SINGLE-STAGE TRANSMITTERS

A. Circuits

The single-stage circuit for whip antenna transmitters on 173 MHz (and 164 MHz) differs slightly from the circuit for loop antennas on those frequencies, in having a low value capacitor from the positive rail to the collector of

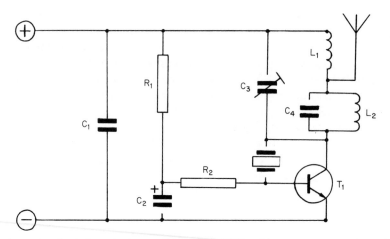

Fig. 3.1. The circuit for a single-stage transmitter, with a whip antenna, on 160–174 MHz. Typical component values are: C_1, 2.2 nF; C_2, 1.5 µF; C_3, 6–25 pF; C_4 8.2 pF; R_1, 470 kΩ; R_2, 3.3 kΩ; T_1, ZT × 314; L_1, 40 s.w.g. enamelled wire, four turns at 2.5 mm; L_2, 40 s.w.g. enamelled wire, 12 turns at 2.5 mm. For 138–160 MHz, C_3 can be omitted, and the trimmer used instead at C_4 to peak the tuning.

The NPN transistor (Fig. 3.1). When carefully tuned, this circuit delivers almost all its output on the 3rd harmonic, with a signal strength 2–3 times that possible in circuits lacking the extra capacitor. The circuit without this capacitor seems to give relatively more third-harmonic output at 150 MHz than at 173 MHz, and not to be greatly improved by adding the capacitor. There is probably still scope for improving the tuning of these circuits: the 150 MHz version is essentially unchanged since it was published in the 1960s (Cochran, 1967).

B. Components

The following description is based on the use of a third-overtone crystal in an SM-1 or HC-45 holder, preferably with a cold-weld or resistance-weld seal. Solder-seal cans can also be used if care is taken not to solder near the seal. If f (in MHz) is the desired operating frequency, then the crystal frequency to order for 138–173 MHz tags is given by $(f - 0.015)/3$ MHz

The Ferranti ZTX 314 transistor is particularly easy to work with, but other small-signal VHF transistors may be used, if possible of the BFY90 type but in small packages (e.g. Ferranti FMMT 2369, Motorola MMT 3904). The other semiconductor component used in single-stage tags, for temperature sensing, is a miniature glass bead thermistor. RS part 151–164 is very suitable, with a 470 kΩ resistance at 25°C.

Carbon composition resistors (⅛ W), such as the RS 141-series, are inexpensive and easy to obtain, but carbon film types from Siegert (RKL 2 10%) or Steatite-Roederstein (RSX 00) are even smaller. The low-value plate ceramic tuning capacitors are low-K long-lead types made by Mullard (supplied by RS, Farnell Electronic Components, and other distributors). FEC also supply the 2.2 nF multilayer ceramic capacitors (143–730), and resin-dipped 1–2.2 μF tantalum capacitors for pulse control. However, for the smallest transmitters it is best to use 1–2.2 nF Redcap ceramic capacitors in the 8101 style, with Corning minidip or cordwood tantalum capacitors in the MD or MZ series for pulse control (e.g. MD2 155 20, MZY 225 20).

The Johansson range of subminiature air-trim capacitors, available from Tekelec Components in Europe, are ideal for final tuning, using part 9402–6 (6–25 pF) in tank circuits and small loops, or 9402–1 (1–5 pF) in large loops. Order a 4092 tuning tool at the same time. The inductor coils are of enamel-coated 40 s.w.g. copper wire, used for transformer and solenoid windings.

Rapid Araldite is an extremely versatile potting substance for small transmitters. It has a good consistency for spreading thinly, and can be made even more fluid by warming if very thin layers are required. It is electrically inert, and sets hard if there is a slight excess of adhesive in the adhesive–hardener mix. On the other hand, an excess of hardener makes it rubbery on setting, which can be useful for transmitters which may need to be repaired or broken up to reclaim components, provided that they will be protected by further layers of harder potting during tag construction.

C. Construction

Success in making transmitters depends above all on making good soldered joints, without over-heating the components. It is well worth pre-tinning all leads, using the fine snipe-nosed pliers or artery clamp to heat shunt the wire as close as possible to the component. Keep the iron clean, and wipe off excess solder on a wet sponge before making a joint, which should be as small as is consistent with a continuous coating round all the leads. The iron should not be applied for more than one or two seconds at a time.

(1) Current and pulse control circuitry

Choose the combination of resistors and pulse capacitor to suit the transistor, the required power output, and the type of cell to be used (Table II). The components in the following description will make a 173.20–173.35 MHz transmitter, weighing 1.2–1.5 g after fairly frugal potting and suitable for a whip antenna. It will draw an average current of 100–200 μA from a lithium/copper-oxide cell, and give about one 25 ms pulse per second.

Table II. Circuit component combinations for single-stage transmitters on 1.35–1.5 V, with about one signal pulse per second. The circuits with the lowest average currents give the longest tag life with each cell type, but have shorter pulse durations and are therefore least easy to detect at long range.

Transistor	"Load" resistor	Bias resistor	Pulse capacitor	Average current
ZTX314	680 kΩ	4.7 kΩ	1.5 μF	80–120 μA
ZTX314	330 kΩ	3.3 kΩ	2.2 μF	130–200 μA
MMT3904	680 kΩ	8.2 kΩ	1.5 μF	60–100 μA
MMT3904	330 kΩ	8.2 kΩ	2.2 μF	120–200 μA
FMMT918	470 kΩ	15 kΩ	2.2 μF	80–120 μA
FMMT2369	330 kΩ	8.2 kΩ	2.2 μF	100–150 μA

Note that variations in transistor quality can cause quite wide variation in circuit current and pulse-rate. Use slightly lower load resistor or pulse capacitor values for individual transmitters in which the current and pulse-rate is too low, and *vice versa*.

You will need:

    A 57.730–57.778 MHz crystal in an SM-1 or HC-45 holder
    A ZTX 314 transistor or equivalent
    A 470 kΩ "load" resistor (RS 141–577)
    A 3.3 kΩ bias resistor (RS 141–319)
    A 1.5 or 2.2 μF pulse capacitor, tantalum bead type (FEC 100 886)
    A 2.2 nF capacitor, plate ceramic type (FEC 143 730)
    An 8.2 pF tank-circuit capacitor, plate ceramic type (RS 125–547)
    A 6–25 pF trimming capacitor (Tekelec 9402–6)
    About 30 cm of 40 s.w.g. enamelled copper wire
    A packet of Rapid Araldite

The stages of construction are shown in Fig. 3.2 (*A–Q*).

A. Bend each crystal lead outwards about 2 mm from its origin and cut the end off flush with the side of the crystal can. Heat-shunt and tin each lead.

B. Hold the crystal in the jaws of the combination pliers "third hand" and tin the side of the can as shown. If the frequency is inscribed on one face of the can, this surface face would be kept free of components, and thus remain readable, if it faced away from you in the diagram. (With a solder-seal crystal, however, the priority will be to avoid soldering at the corner which has the seal.)

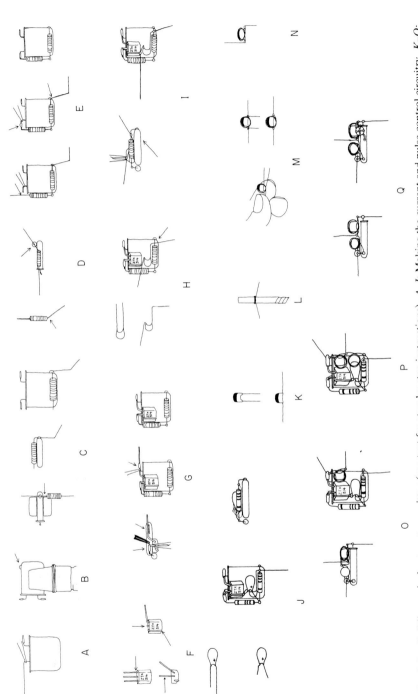

Fig. 3.2. Building a single-stage transmitter (see text for step-by-step instructions). *A–J*: Making the current and pulse control circuitry. *K–Q*: Adding the tuning circuitry.

C. Solder one lead of the 470 kΩ "load" resistor to the crystal can, 3–4 mm from the body of the resistor. Hold the lead in pliers close to the resistor body, to prevent it breaking away, and bend the lead so that the resistor lies flat across the crystal can. Cut off the other lead 2–3 mm from the body of the resistor.
D. Bend one lead of the 3.3 kΩ bias resistor about 30°, as shown. Holding the body of the resistor against the side of the crystal can, solder the bent lead to the free lead of the load resistor. The join should not be closer than 1 mm to the crystal can.
E. Bend the free lead of the bias resistor to lie alongside the crystal lead and solder them together, heat-shunting the crystal lead. Cut off the free ends of the resistor leads.
F. Bend the transistor leads as shown. Cut off the base (central) lead about 2 mm beyond the point where it crosses the plastic case (see G), and trim the emitter lead flush with the edge of the plastic case.
G. The crystal lead which joins to the bias resistor should be held in the artery clamp as a heat shunt, with the artery clamp gripped in the "third hand". Hold the base transistor lead in the snipe-nosed pliers to tin it, then solder it to the bias-crystal joint, and release the artery-clamp heat shunt. Bend the collector lead to lie alongside the other crystal lead, grip both wires together to heat shunt, solder them, and trim the excess off the collector lead. Make sure the transistor leads are clear of the crystal case.
H. Bend the 2.2 nF capacitor leads as shown, one through a right angle in a horizontal plane about 4 mm from the capacitor body, and the other through 180° in a vertical plane. Thread the right-angled lead between the crystal can and the load resistor lead, press it upwards against the load resistor lead, and solder these leads together. The capacitor should fit neatly into the gap between the transistor and the load resistor, with its other lead not closer than 1 mm to the crystal can.
I. Solder the free capacitor lead to the emitter lead of the transistor, which should be heat-shunted. Do *not* trim the capacitor leads, which will serve as positive and negative leads in the finished transmitter.
J. The pulse capacitor is polarized, with its positive lead the longest. Place the pulse capacitor over the 2.2 nF capacitor, bending its leads as necessary to bring the positive one against the junction of the load and bias resistors, and the negative lead against the junction of the transistor emitter and 2.2 nF capacitor leads. Solder the pulse capacitor leads at these points, making sure that they remain 1 mm clear of the crystal can, and heat-shunting the emitter lead. Trim the excess off the pulse capacitor leads.

(2) Tuning circuitry

K. Bend the 8.2 pF tank-circuit capacitor leads outwards in opposite directions. Crush any ceramic off and tin them, so that the inductor coil leads may be soldered as close to the body of the capacitor as possible.
L. Wind 12 turns of the 40 s.w.g. wire round the ⅛ in (3 mm) drill bit, crossing the turns over each other to hold them together and keep this inductor as compact as possible. Leave the ends about 30 mm long, and tin them over a 5–10 mm length starting within 1 mm of the coil: hold the wires aginst the soldering iron tip, which must be very hot, and allow fresh solder to flow over them.
M. Remove the coil from the drill bit, pushing with your finger and thumb nails and being careful to keep the coil neat. If you wish, you can then spray the coil with lacquer to help it hold its shape. Place the coil against one face of the 8.2 pF tank-circuit capacitor, holding it in place between your forefinger and thumb. Wind the free ends of the coil several times around the trimmer leads, as close to the body of the capacitor as possible, solder, and trim. These joints should be as small as possible, but make absolutely sure that they are sound, and that you have not simply covered an enamelled surface with solder. Poor connections here are one of the most common sources of transmitter failure, sometimes after several days of satisfactory operation.
N. Bend one tank-circuit lead at its emergence, at right angles to the plane of the inductor coil, and the other in the opposite direction about 2 mm from its emergence, being careful not to disturb the coil.
O. Heat-shunting the junction of the collector and crystal leads, solder the tank circuit to them so that the capacitor is parallel with the crystal can, with its leads at least 1 mm clear of the can.
P. Wind a four turn coil on the drill bit and tin its ends, as in *L*. Place this inductor over the end of the pulse capacitor, and wind one end several times round the nearest lead of the tank-circuit capacitor, as close as possible to the lead's emergence. This tank-circuit lead, at which the inductor coils join, is the antenna lead. Keep the coil connections as close to the body of the capacitor as possible: otherwise the fine inductor wire is easily broken when the antenna is fastened to the finished transmitter. Solder the other end of the four-turn coil to the crystal can at its junction with the load resistor. Trim off the free ends of the inductor wire.
Q. Hold the trimmer against the side of the crystal can, and solder one lead to the can at its junction with the load resistor. This lead is then bent to ensure that the capacitor is clear of the can and the tank circuit. The

tank-circuit lead which extends beyond the collector–crystal junction is bent up to meet the other trimmer lead. The trimmer is soldered to this lead, using a heat-shunt as shown to prevent the tank circuit being disconnected from the collector–crystal junction. The trimmer can also be soldered between this end of the tank circuit and the crystal can in other positions, to keep it accessible for final tuning when the tag is completed.

D. Using other components

Transmitter size can be reduced, with some increase in cost, by using Motorola MMT3904 transistors. In this case, the bias resistor should be 8.2 kΩ and can lie along the edge of the crystal can's top face, rather than at the side. The body of the transistor fits neatly between the crystal leads, with its green face inwards and the emitter lead pointing upwards. The collector and base leads are each in turn held against a crystal lead, using snipe-nosed pliers or the artery clamp to grasp and heat-shunt both components as they are soldered together (Fig. 3.3). If necessary, a dab of cyanoacrylate glue ("Superglue") will hold the body of the transistor in place against the crystal can during this rather fiddly operation.

These transistors are all equivalent to the BFY-90, a comparatively bulky component in a metal can, which can be used as a last resort. The surface-mount Ferranti FMMT2369 and FMMT 918 types are also suitable, and cheap, but very fiddly.

If cost-saving rather than peak performance is the main aim, the trimmer can be replaced by a fixed value plate ceramic capacitor like that in the tank circuit, a value between 10 and 22 pF being chosen by trial and error to give the strongest signal (15 pF is a good compromise value).

To build the smallest transmitters, use carbon-film resistors, Redcap 2.2 nF capacitors and the smallest Corning tantalum pulse capacitors. After much practice the weight can be brought down to about 600 mg, if the layout of these components is altered to make the assembly as compact as possible.

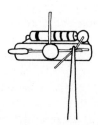

Fig. 3.3. Using a MMT transistor in a single-stage transmitter.

E. Trouble-shooting

Before you connect the completed circuit to the battery and test meter, it is wise to use the resistance scale of the multimeter to check that there is no short circuit in the transmitter. Since resistance is measured by sending a small current through the tested components, the meter should either pulse (as the transmitter operates) or register a high resistance of several hundred kΩ depending on the polarity of the test leads. If there is a steady negligible resistance whichever way you connect the meter, there is a short circuit in the transmitter. Check that the components are correctly connected, especially the transistor, and that the negative lead in particular is not touching the crystal can.

If there is no obvious short circuit, connect the transmitter in series with the battery and multimeter. Provided that your lead polarity is correct, you can expect a current which pulses between about 80 and 250 μA, but there are other possibilities:

1. The meter needle rises sharply off the scale. Disconnect immediately and check again for short circuits.
2. The meter needle rises to between 100 and 250 μA but does not pulse. The tuning may be incorrect. First check that turning the trimmer through a complete revolution does not produce a pulse. Secondly, look to see that inductor leads are not bared and thereby causing short circuits between coils in the tuning circuitry, especially in the tank circuit. A third check for tuning problems is to replace the tank circuit, also trying capacitors of higher and lower values than 8.2 pF.

    Before you remove anything, however, it is worth checking (i) that both leads of the pulse capacitor are properly connected, and (ii) whether the transmitter pulses when a parallel resistor of 330–1000 kΩ is used to decrease the load resistance. If so, and the steady current without the reduced resistance was above 100 μA then the equivalent series resistance of the crystal may have increased through heat damage (assuming that the crystal was checked before assembly). If the current was low to start with, then the transistor gain is poor: you can either replace it or use a lower load resistance.

    If all these steps fail, then the crystal is probably faulty, and you will have to disconnect a number of connections even to test it. Remember to shunt heat-sensitive components while you do this.
3. The circuit current is steady and between 10 and 80 μA. You either have a damaged load resistor, or have damaged the transistor (assuming its gain was adequate before assembly). Check the resistor with the multimeter and inspect it for cracks, especially near the end of its casing. If the resistor is undamaged, replace the transistor.

4. The meter needle moves almost imperceptibly up the scale, and then stops. You may have damaged one of the resistors, but the most likely fault is a poor connection somewhere in the tuning circuitry. Check for an open circuit across the tank circuit and the smaller coil.
5. There is no reaction from the meter at all. Your transmitter has an open circuit, quite possibly because the positive lead is not properly connected. This is a common cause of intermittent operation, especially if the positive lead has been overheated when connecting cells during tag assembly.

F. Tuning

If your transmitter pulses satisfactorily, it is a good idea to peak the tuning before potting, just in case the trimmer gets glued up. Solder an antenna to the end of the lead which joins the inductor coils, and a cell to the transmitter's positive and negative leads. Keep the cell leads as short as possible and do not connect the meter, because this would affect the tuning, as would touching the transmitter assembly except with the insulated tuning tool: you may need to hold the transmitter down with modelling clay (e.g. Plasticine) while you tune it. Putting the receiver near enough to the transmitter to hear the signal clearly, insert the tip of the tuning tool into the rectangular slot in the trimmer and turn gently until the signal becomes strongest. The pulse will probably slow slightly as you approach this point. If the receiver has no signal strength meter, use earphones or an earplug to help detect the changes in signal strength.

G. Potting

Before you pot the transmitter, write its frequency on a "flag" of insulating tape wrapped round the positive lead (Fig. 3.4A). If you neglect to do this, and potting makes the label on the crystal can unreadable, you will waste a lot of time using your receiver to find the frequency again.

Mix a small quantity of Rapid Araldite adhesive with a similar amount of hardener. Using the small screwdriver, give the transmitter a thin coating, taking care neither to disturb the inductor coils nor to cover the centre of the trimmer (Fig. 3.4B). You will find the transmitter easy to pot if you clamp the positive lead in the artery forceps.

The Araldite sets within about 15 min, at which time the tuning tool must be used to turn the trimmer slightly. If you wait more than about 30 min before checking that this capacitor is free, it may be firmly stuck. Applying

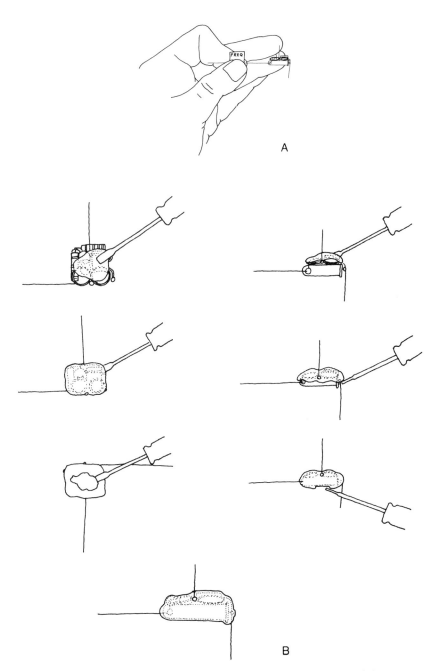

Fig. 3.4. A single-stage transmitter, with its frequency label, ready for potting (*A*), and during potting with Rapid Araldite (*B*).

heat round the edges with the soldering iron sometimes loosens stuck trimmers, but note that epoxy resins give off toxic fumes when heated in this way.

H. Other frequencies

The same circuit works well at the slightly lower frequency of 164 MHz. Moreover, a similar transmitter can be built with a whip antenna on 104 MHz. The 104 MHz circuit has 16 turns in the tank circuit on a 2.2 pF capacitor, and seven turns on the second coil, with a 22 pF fixed-value capacitor from to the positive terminal (earth) to the transistor collector. A 6–25 pF trimmer is soldered across the leads of the 2.2 pF capacitor (Fig. 3.5), because the tank circuit needs very careful tuning at this frequency.

At 150 MHz a good third harmonic can be obtained without a capacitor to earth the collector, using a 14-turn coil on a 2.2 pF capacitor to form the tank circuit, with the trimmer across it as in Fig. 3.5, and four turns in the second coil. Care is needed when tuning this circuit. With very low tank-circuit capacitance it will not pulse. As the trimmer is turned to increase the capacitance, a prolonged "weepy" pulse is obtained, which reduces in length to a short, sharp "peep". If the trimmer is turned even further the signal may become clipped, but may also become a continuous scream, due to a feedback effect, and rapidly burn out the transistor. This overloading is especially likely with ZTX 314 transmitters powered by lithium/copper-oxide cells, which are nominally 1.5 V but give as much as 2.2 V when first used. Since potting has the same effect as increasing the tank-circuit capacitance by 1–2 pF, the final tuning of transmitters with these cells should always be done when the tag is complete.

Another option at 150 MHz is to use a fixed value 12 pF capacitor and 14-turn inductor in the tank circuit. This reduces the risk of feedback problems and is also much cheaper than using a trimmer.

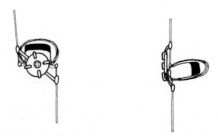

Fig. 3.5. Making a tank circuit with a 6–25 pF trimming capacitor.

I. Loop antennas

In transmitters for use with loop antennas, construction finishes at stage O. On 173 MHz the value of the fixed tank-circuit capacitor, with a 12-turn inductor, is increased to 18 pF; 22–25 pF is appropriate at 150 MHz. The antenna loop is connected from the antenna lead of the tank circuit to the positive terminal, replacing the second inductor coil of transmitters for whip antennas. Tune the loop to resonance by adding a capacitor across it. For small tags with loops of less than about 10 cm circumference, you can build a 5–25 pF trimming capacitor into the transmitter, by soldering one end to the crystal can and the other to the tank circuit antenna lead. With larger loops the tuning capacitor may be added during tag assembly, either as a 1–5 pF trimmer or as a fixed value capacitor selected to give the strongest signal.

J. Sensor circuitry

If a mercury activity switch is to be used, two pulse capacitors are fitted at stage J. For instance, two 1.0 μF capacitors might be used instead of a single 2.2 μF component. One capacitor is connected from the emitter to the junction of the load and bias resistors, as in J, but only the positive lead of the other is connected (to the resistor junction). In the finished tag, the mercury switch is connected from the remaining lead of this second capacitor to the negative lead of the transmitter. Only 1 μF is in circuit with the switch open, giving a rapid pulse, but when the switch closes the second capacitor is connected in parallel with the first, to give a total 2.0 μF, and the pulse rate is substantially reduced. Capacitance values must be selected to give the desired combination of pulse rates, taking account of whether the transmitter is for high or low current operation.

A miniature glass bead thermistor, with a resistance of about 470 kΩ at 25°C (e.g. RS 151–164), can completely replace the load resistor in these transmitters. In a carefully tuned tag, this can enable temperatures to be estimated with an accuracy of about 1°C, but with a substantially reduced tag life (see Chapter 3.IV.E). The pulse rate of each tag will differ slightly at the same temperature, so each will need to be calibrated separately.

If a reduction in accuracy is acceptable to prolong tag life, a 1 MΩ load resistor can be used, leaving a lead extending from the load-bias junction at stage D. The thermistor is later connected between this lead and the transmitter's positive lead, with a 100–470 kΩ resistor in series: the two components thus parallel the 1 MΩ load resistor. A 330 KΩ resistor in series with the RS 151–164 in squirrel collars is adequate for determining drey uses, with no more than 30% reduction in cell life.

III. A TWO-STAGE TRANSMITTER

A. Construction

These transmitters are constructed on a 10 × 12 mm PCB. The track layout is shown in Fig. 3.6 and the finished transmitter in Plate 3 (lower centre and right). If you get boards made for you, use a substrate not appreciably thicker than 1 mm. The holes should be 0.6–0.8 mm diameter, apart from the two large inductor holes (Figure 3.6*I,J*), which should be 1 mm across. Biotrack (see Appendix I for address) can provide kits of boards and components for these transmitters. Check that there are no shorts between tracks on the board, especially at the edges where cutting machinery may have smeared metal across a gap: scrape the edge of the insulation with a knife to remove any unseen metal film.

Make sure all components are well soldered to the board. Holding the inserted component in position on the board with the fingers of one hand, arrange the solder so that its end touches against one lead's emergence on the other side of the board, without holding the solder. This keeps one hand free for the soldering iron, and is easiest if you get a short length of fine solder to stand out horizontally from its container. When you touch the solder with the iron from the other side of the component lead, solder should flow smoothly over the track all round the lead. The lead can then be cut short where it leaves the solder. Remember to avoid overheating, by applying the iron for no more than 1–2 s at a time, but make sure that the joints have concave, shiny surfaces which blend smoothly with the track and the lead.

You will need:
  A 57.730–57.778 MHz crystal in an SM-1 or HC-45 holder
  2 ZTX 314 transistors or equivalents (see Section II)
  A 330–470 kΩ first-stage "load" resistor (RS 141 series)
  A 1.5 kΩ first-stage bias resistor (RS 141–274)
  A 3.3–4.7 μF sub-miniature pulse capacitor (Corning MZY series)
  A high-μ toroid, 4–6 mm external radius by 1.5–2 mm thick (e.g. Neosid 35–507–34)
  A 15 pF capacitor (e.g. FEC 104 880)
  A 15 kΩ second-stage bias resistor (RS 141–397)
  A 2.2 nF capacitor (FEC 143 730)
  2 6–25 pF tuning capacitors (Tekelec 9402–6)
  A tuning tool (Tekelec 4192)
  About 20 cm of 34/35 s.w.g. enamelled copper wire (RS 357–766)
  About 10 cm of 22 s.w.g. enamelled copper wire (RS 357–924)

3. Building Transmitters

For 2.7–3.5 V supplies, with an average current of 200–300 μA, use a 470 kΩ first-stage load resistor and a 3.3–4.7 μF pulse capacitor. The pulse rate will be greatest with the lower-value capacitor. The gain of ZTX 314 transistors is quite variable, and those with really low gain may require a 330 kΩ load resistor to give a reasonable pulse rate at 2.7 V. The transmitter can also be built for 1.5 V operation, with a 220 kΩ load resistor and 1.5–2.2 μF pulse capacitor. As with the single-stage transmitter, the first-stage load resistor can be replaced by a thermistor form temperature sensing, or the pulse capacitance split for use with a posture-sensing mercury switch.

The stages of construction are shown in Fig. 3.6.

A. Bend the first-stage load resistor leads to fit through the holes indicated. Solder the junction on the larger track, but leave the second junction until the other component leads are in place there. In this way, there is no risk of blocking the second hole while soldering the lead through the first.

B. Insert the ZTX 314 second-stage transistor and solder the outer (emitter and collector) leads. Cut off the surplus central (base) lead about 1 mm beyond its emergence, but do not solder it yet. If you use a different second-stage transistor, make sure that the collector is at the corner of the board, with the base in the middle and the emitter third in line.

C. Bend the 1.5 kΩ resistor lead and insert as shown, but solder only the junction on the longer track.

D. Insert the positive lead of the pulse capacitor through the same hole as the free load resistor lead, but do not solder it. The capacitor lead should be thin enough to fit with the second lead. If not, the pulse capacitor lead can be soldered to the long vertical lead of the 1.5 kΩ resistor. This resistor lead also makes a convenient junction for the second pulse capacitor used in posture sensing.

E. Insert the first-stage ZTX 314 transistor, passing the free (negative) pulse capacitor lead together with the collector lead through the track at the side of the board. Solder the central (collector) and inner (base) leads, but not the emitter.

F. Insert the crystal, preferably with the frequency label outermost so that you can read it later if required, and solder the leads. Be careful that the crystal neither damages the 1.5 kΩ resistor, nor touches its leads. With the crystal in place, you can adjust the position of the pulse capacitor so that it sits neatly between the transistor and the crystal. Check that the capacitor leads do not contact any other leads or the crystal can, and then solder all the junctions left free in stages *A, C, D* and *E except* the base of the second-stage transistor.

64   Wildlife Radio Tagging

G. Wind 12 turns of 34/35 s.w.g. enamelled copper wire on the toroid, and tin the free ends. Insert the ends as shown, but do not solder them yet.

H. Insert one tuning capacitor lead through the corner hole, so that the toroid is positioned neatly between it and the first-stage transistor. Solder the capacitor lead, and the toroid lead on the same track. Insert a 10–15 mm resistor lead offcut through the hole in the adjacent track, solder this lead and the other toroid lead. Bend the free wire to contact the free lead of the trimmer and solder them together.

I. Wind 3.5 turns of 20 s.w.g. wire round a 2.5 mm diameter drill bit, and shape as shown. The free ends should be cut off about 2 mm beyond the periphery of the coil, stripped and tinned. Insert the coil and solder as shown. This stage is omitted in transmitters which are for use with loop antennas.

J. Wind 4.5 turns of 20 s.w.g. wire on the same drill bit. Shape, strip and tin the coil as in *I*, before soldering it in place. Solder one end of a

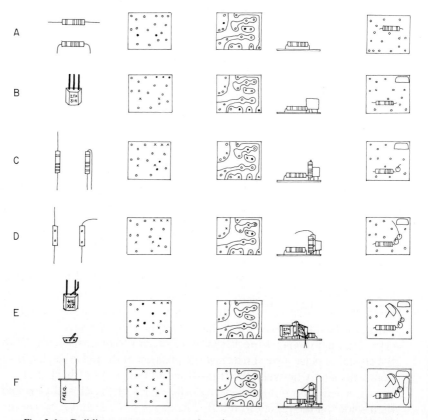

Fig. 3.6. Building a two-stage transmitter (see text for step by step instructions).

20–30 mm length of wire to the free end of this coil, and to the other coil (if present): the other end of this wire stands free as the tuning lead.

K. Attach the remaining three components to the lower surface of the board. The 15 pF capacitor has one lead bent to the side, before the other lead is soldered to the base of the second-stage transistor. Use as little solder as possible, to avoid any risk of it spreading across to the adjacent tracks. Since the track has not been tinned at this point, it will absorb a slight excess of solder. The other capacitor lead should be soldered along the line of the other track, and the excess trimmed from both leads.

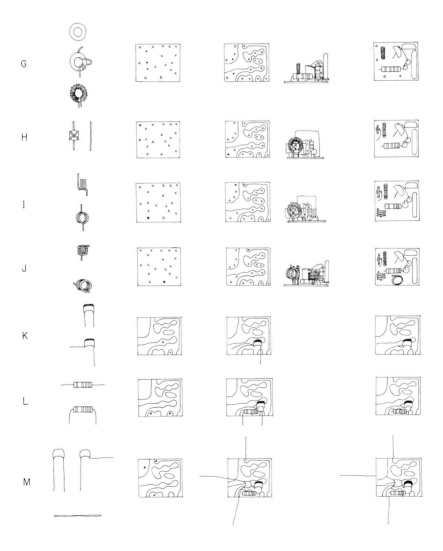

L. Bend the leads of the 15 kΩ resistor (second-stage bias) close to the body of the component. Solder one lead to the capacitor–base junction, and the other to the junction of the two transistor emitters, again using as little solder as is necessary for sound joins. Trim both leads.

M. Bend the leads of the 2.2 nF decoupling capacitor as shown. Solder one lead soundly to the track at the end of the board: the free end is the positive lead. Solder the other lead to the 15 kΩ–emitters junction: the free end is the negative lead. Check that neither of the decoupling capacitor leads come within 1 mm of the 15 pF capacitor lead and track. Finally, solder a 20–30 mm length of insulated wire to the track at the bottom of the second large inductor coil: this is the antenna lead.

B. Tuning

The transmitter is now ready for preliminary tuning. If a whip antenna will be used, this should be attached to the antenna lead, and the second tuning capacitor connected between the antenna and tuning leads. If the tag will have a loop antenna, this is connected between the positive and tuning leads. The tuning trimmer also connects between the positive and antenna leads, across the loop.

The tuning capacitor on the board acts with the toroid inductor to resonate the first stage at the crystal's first harmonic. For transmitters between 138 and 174 MHz, the circuit should oscillate over a wide range of capacitor settings and the pulse rate slow appreciably at resonance. The second stage acts as a frequency multiplier, such that tuning for the loudest signal in this frequency band will peak the third harmonic. The adjustment is very sensitive, and is best done by attaching the second variable capacitor and peaking the tuning in the final stages of potting the tag (Chapter 4). At the lower frequencies a 10 pF capacitor may be needed in parallel with this 5–25 pF tuning capacitor.

At higher frequencies the number of turns on the toroid can be reduced. The frequency multiplier stage will also resonate well on the fourth harmonic if the second-stage tuning capacitance is high enough. For example, a good 216 MHz signal can be obtained from a 54 MHz crystal.

The completed transmitter weighs about 2 g, and for best results should be sealed in a 2 cm length of heat-shrink tubing with an adhesive or meltable lining (e.g. Raychem 3/8 SCL), so that the inductors remain in air instead of being embedded in potting. The tubing can be squeezed against itself with pliers while it is warm, to give a watertight seal round the leads and at the other end of the board (Plate 3, lower right).

4
Tag Assembly

Radio tags, containing basic transmitter circuits connected to power sources and antennas, are completed by encapsulating them in a shape suitable for attachment to the study animal. Appropriate mounting materials are also built into the tag, and provision is made for the circuit to be broken after assembly and testing if the tag is not for immediate use.

I. POWER SOURCES

A. Cell lead attachment

Since it is easy to overheat cells by soldering leads directly to them, it is usually best to obtain cells with solder tags. If small cells are overheated locally, in the area where a lead is attached, they may be damaged such that their capacity diminishes and the tag's life is thereby reduced. More severe overheating can lead to total failure of the cell, due to the establishment of internal short circuits, or even to the cell exploding. However, leads can be soldered to even the smallest cells using flux-cooled soldering. You will need fine soldering equipment, pliers set up in a vice as a third hand (see Chapter 3.I), insulating tape (e.g. RS 511–954), a pair of snipe-nosed pliers, tinned copper wire of an appropriate thickness for the leads, and a few ml of 10% phosphoric acid. You should also wear protective goggles or a mask.

The third-hand pliers are used as a heat-sink, to prevent severe overheating. To heat-sink mercury or silver oxide button cells effectively, without short-circuiting them between the jaws of the pliers, one of the opposing jaws should be wrapped in insulating tape. It is best to attach the first lead to the curved side of the cell casing, which is the anode of mercury or silver cells. Put the cell in the pliers with its casing gripped against their uninsulated face (Fig. 4.1*A*), being careful that the opposing electrode does not also contact an uninsulated part of the pliers while you do this. Then place a 1 mm diameter drop of 10% phosphoric acid on the cell's surface where the lead will be attached, using a capillary tube or a loop of wire dipped in the acid (Fig. 4.1*B*). The acid serves as a flux and to prevent local overheating. Hold the lead with one hand against the cell's surface in this spot of flux. The iron is held in the other hand, wiped clean on the soldering stand's damp

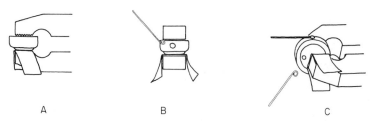

Fig. 4.1. Flux-cooled soldering for small cells. The cell is gripped in pliers, with the negative terminal against a layer of insulating tape (*A*). A fine wire loop puts a drop of 10% phosphoric acid on the casing (*B*), and the positive lead is soldered. The process is repeated for the negative lead (*C*).

sponge and tinned with a very slight excess of solder, then immediately wiped quickly against the lead and the cell surface where these meet. The iron should be used with a rubbing rather than a dabbing action, and should not contact the cell for more than half a second. There will be a brief hiss as flux evaporates, removing excess heat from the cell surface as latent heat of evaporation. You should then remove the cell from the pliers and test its temperature against your lip. It will not have been damaged by slight warming, but should be rejected if it is uncomfortably hot. The soldered joint should then be thoroughly wiped to remove any traces of phosphoric acid, which might eventually corrode the wire lead. The joint to small mercury and silver cells should be no more than about 2 mm in diameter, perhaps somewhat larger to ensure firm lead attachment to large cells.

In joining the second lead to the central electrode (Fig. 4.1*C*), again take care that the cell is not tilted such that the edge of the casing or the first lead makes a short circuit against bare metal. If the first lead was for any reason soldered to the flat base of the casing, rather than to its curved side, beware of any projecting point of solder which might pierce the insulating tape to cause a short circuit while the other lead is being attached.

An alternative lead attachment technique for very small cells is the use of conducting paint or epoxy. The tinned lead wire is wound in a tight flat spiral, to ensure a large area of contact with the cleaned cell surface, and adhesive or paint is smeared on both surfaces before they are brought together. The joint can then be clamped to harden in an insulated holder, protected if necessary from glue or paint with a sheet of polythene or other non-stick plastic, and taking care that the lead from a central electrode is not pressed against the cell's outer casing. This is a slower technique than flux-cooled soldering, and tends to produce a weaker join, with a greater weight of wire and attachment materials, but is less likely to damage the cell.

Small button cells can sometimes be obtained with solder tags already attached by spot-welding. However, spot-welding often reduces the

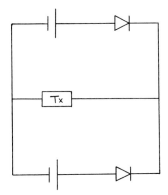

Fig. 4.2. Parallel cells in transmitters should be isolated with germanium diodes.

capacity of the smallest cells, and mercury cell manufacturers in Britain have now refused to attach these tags because of the high wastage rate. Skilful flux-cooled soldering may well be the best way to attach leads to button cells. After some practice, cells can be soldered directly to transmitter leads with no appreciable warming of the cell, even without pliers as a heat-sink.

B. Diode isolation

Sometimes two cells must be used in parallel to meet tag shape or life requirements. For instance, if a tag with a maximum diameter of 8 mm must have a life of 3 months, you could connect a low power transmitter to two 85 mA h RM13 cells in line. Note, however, that putting two cells in parallel does not necessarily double the life of the tag. If the cells discharge unevenly, such that one is exhausted well before the other, the second cell will then discharge quickly through the circuit formed by the exhausted cell. This can be prevented by isolating each cell with a germanium diode (e.g. RS 261 952), as in Fig. 4.2. Silicon diodes should not be used because they have a higher voltage drop (0.7 V) than germanium diodes (0.3 V). Since the diodes increase tag size, they are hardly worth using in the smallest packages, but they fit easily in the space between most lithium cells and should certainly be included in the larger tags. Make sure that the tags operate adequately with the reduced voltage.

C. Solar panels

A germanium diode should also be used to isolate the solar panels in tags which include a rechargable cell (Fig. 4.3). Without a diode, power will be

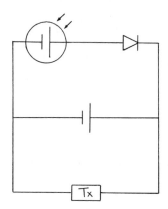

Fig. 4.3. Solar-powered tags with rechargeable cells should have a germanium diode to prevent current drain through the solar panel when there is little light.

drawn by the panels in the dark. Unfortunately, even the 0.3 V voltage drop of germanium diodes is enough to require a slightly larger solar panel area than without a diode.

A particular problem with these tags is the inadequate charging of their cells during short winter days at high latitudes. Although it may not matter if a tag stops transmitting towards the end of the night, rapid discharge of the NiCad cell will minimize the chance of finding dead animals if their tags are poorly lit. To be on the safe side at high latitudes, a tag should charge enough in typical habitat on a dull day to run for at least four days unlit.

To protect solar panels from damage by the animal or its surroundings, they should be encapsulated with an overlying layer of hard transparent plastic. If the tags are mounted on birds, there should be enough stiff fabric or thin plastic sheeting around the tag to prevent feathers overlapping the panels.

II. ANTENNAS

A. Whip antennas

Transmitter whip antennas are usually single or multiple strands of nickel, stainless steel or other high-tensile alloy. Whip antennas used on the smallest birds, bats and reptiles are typically 0.020–0.025 mm (0.008–0.010 in) nickel, as used for guitar strings. Plain guitar wire is readily available in thicknesses up to 0.055 mm (0.022 in) and makes good antennas for backpacks on some birds, although dental-brace wire and two-strand or three-strand steel wire may also be used.

However, the thicker guitar wire is too inflexible for whip antennas on most mammals and reptiles, or on hole-nesting birds. A more suitable material for these species is nylon-coated multistrand stainless steel fishing trace, the finest of which has a breaking strain of about 10 kg (25 lb). Fishing trace with at least 25 kg (60 lb) breaking strain makes good antenna material for birds which are liable to bend guitar wire in their beaks, such as corvids or raptors, and for long-life tags in corrosive environments (e.g. on seabirds). Multistrand steel cable for bicycle gears or brakes may be used on the most powerful birds, with a coating of tough meltable or adhesive-lined heat-shrink tubing. Nevertheless, determined birds with powerful beaks, such as some parrots, may eventually destroy any antenna, if not the tag itself.

The whip antenna wire on leg-mounted or tail-mounted tags must be light yet strong, and tolerate a tremendous amount of flexing. The best material seems to be 49-strand steel cable of a type used for control cables in light aircraft. This can be coated with heat-shrink tubing to prevent "fraying", and to reduce flexibility: if these antennas are too flexible, they whip round and snag on branches or wire fences.

Nickel guitar wire takes solder readily, but this is not always true of multistrand stainless steel. After stripping a short length of nylon coating from the end, the bare wire should be dipped in 10% phosphoric acid flux. It usually then tins without too much difficulty. A strong pair of clippers is needed if these materials are to be cut without damage to the blades, especially for the brake or gear cable. To prevent the strands of heavy cable from being forced apart when they are cut, heat the cable at the chosen point and allow solder to flow into it. The cable then cuts cleanly and is pre-tinned.

One of the most common problems with whip antennas is breakage, through metal fatigue and/or corrosion, at the point of emergence from the potting material. There are three main ways to prevent the bending strain being concentrated at this point. The oldest approach is to set a 2–3 cm (1 in) long compression spring into the potting, so that the antenna passes through the spring as it emerges (Fig. 4.4A). Springs from ballpoint pens are ideal for the smaller antennas. The second approach is to form a slender cone of silicone sealant, such as that used for caulking baths, around the antenna as it emerges. However, if the silicone compound gives off acetic acid on curing (e.g. RS 555–588) it can corrode and weaken metal antennas. This seems not to be a problem within the cell-life of small tags, but guitar wire antennas intended to last more than six months should have their bases coated with heat-shrink or other tubing to prevent this corrosion (Fig. 4.4B). Another way of spreading the strain at antenna emergence points is to use concentric layers of flexible heat-shrink tubing (Fig. 4.4C). The finest tubing extends about half-way along the antenna, the next finest for one-third of the way, and so on for three or four layers. This approach works well for fishing

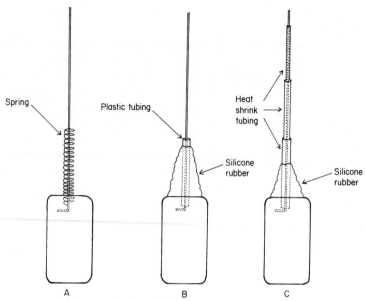

Fig. 4.4. Whip antennas can be supported at their emergence with springs (*A*), with a cone of silicone rubber (*B*), or with layers of heat-shrink tubing.

trace antennas, but guitar wire is too stiff to receive much support from the tubing unless a silicone cone is used as well.

B. Loop antennas

Although whip antennas are sometimes used for very lightweight transmitters at 104 MHz and lower frequencies, tags on these frequencies most often contain integral multiturn antennas wound on ferritic cores. To keep weight to a minimum, cores should be of materials with the highest possible relative permeability to the electromagnetic flux. Since potting can affect the tuning of these antennas quite markedly, provision should be made for a final adjustment when the tag is complete. This can be done by covering the screw slug with wax prior to potting, and leaving one end exposed so that its position in the coil can be adjusted after the potting has set. This system is also used for final adjustment of the output stage on some commercial transmitters at higher frequencies. A more elegant method, however, is to use a miniature trimming capacitor, which is set in the potting with its tuning slot exposed. After peaking the tuning, the access is sealed with potting. Alternatively, if retuning may be necessary and the site is unlikely to be chewed, silicone sealant or other soft material may be used as a covering.

When a 104 MHz tag design is to be built in large numbers, final tuning can be avoided by using fixed value inductive and capacitive components, with an antenna built on to the circuit board. In one design of this type, a spiral track on one side of the board provides a lightweight alternative to a cored antenna, at some cost in radiating efficiency.

Integral tuned-loops are also used in the 138–174 MHz band, typically on implants or where external antennas are likely to be damaged. Although tuned-loop collars of fine wire are often chosen for small rodents, an alternative is to wind a length of enamelled 0.22 mm (34 s.w.g.) copper wire horizontally twice round the edges of the transmitter and cell before final potting. A subminiature trimming capacitor (e.g. Tekelec/Johansson AT9402–6) is then used to tune this small double loop, as it would have been with the larger single collar loop. The integral antenna gives only 30–50% the range of the larger loop on a 25 g mammal, but transmission will continue if social grooming or a predator cuts the collar.

Tuned collar loops on small mammals may be made from multistrand copper wire with a thick coating of PTFE or other very tough plastic. In one design, the collar is made as a continuous loop, with the wire folded in a stretched "figure 8" inside a 1–1.5 cm length of heat-shrink tubing at the top of the collar (Fig. 4.5). Once slipped over the animal's head, the ends of the "figure 8" are teased out from the heat-shrink to reduce the collar's circumference, before the tubing is set with a soldering iron or heated penknife blade (Chapter 5).

Some people dislike fiddling with the bulky "sliding-8" attachment, and prefer their small-mammal tags to have an open loop. A 1 cm length of heat-shrink tubing is slipped over the wire on one side of the collar, and the insulation on both sides is stripped to the circumference of the animal's neck (Fig. 4.6A). The wires are then twisted together, and preferably soldered, before the heat-shrink tubing is warmed over the join (Figure 4.6B). This design benefits by starting to transmit only when the loop is joined. A

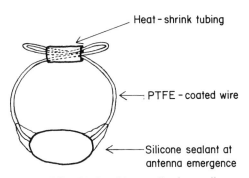

Fig. 4.5. A "sliding-8" closed-loop collar for small mammals.

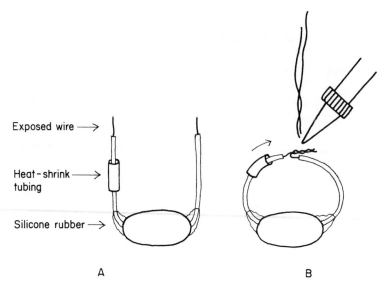

Fig. 4.6. An open-loop collar for small mammals. Insulation is removed from the antenna ends (*A*), which are then twisted together and soldered round the animal's neck (*B*) before being covered with heat-shrink tubing. Use a strip of paper under the wire to protect the animal while you make the join.

separate connection must be used to start closed-loop tags, but they have the most efficient tuning. Open-loop collars should be pre-tuned with their wires twisted together at the average neck circumference, and best results on much larger or smaller animals will only be obtained if the tags are re-tuned after attachment.

Open-loop wire collars can also be attached by setting a metal crimp over the bared wire ends, a method which is particular suitable for tagging mammals weighing 100–400 g. These larger collars should be made with multistrand brass cable, such as that used for hanging pictures, because radiation efficiency increases with antenna thickness at a constant loop circumference. The thick collar wire must be well anchored inside the potting, for instance by soldering it to thin brass strips which are bent under the cell on one side and the transmitter on the other. The wire should also be covered with heat-shrink tubing or other coating to protect the animal's neck (Fig. 4.7). On all the wire loop collars, but especially with the thicker wire, the points where the collar emerges from the potting should be protected from stress and moisture penetration with a cone of silicone caulking.

Larger tuned-loop collars are best made with brass strip. Brass of 0.15 mm thickness can be used for the smaller collars, with a width of 4–8 mm, depending on the shape of the animal's neck (the wider strip being

4. Tag Assembly

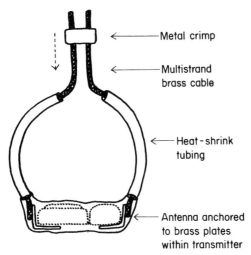

Fig. 4.7. Crimp closure of an open-loop collar. The metal crimp is slid down to the required collar circumference, and then closed tightly. The protruding ends of the collar are trimmed, and may be soldered together beyond the crimp to maintain a good electrical connection.

the more efficient radiator). For the largest sizes, 10 mm wide strips of brass, 0.4–1.0 mm thick, are suitable. Use thick brass for fixed circumference loops, to increase rigidity and radiation efficiency, and the thinner brass strip to give greater flexibility in collars with an adjustable circumference.

Closed-loop brass strip collars must have an internal circumference as great as the head of the largest animal to be tagged. One way of fixing the collar in place is to screw a gusset of moulded plastic (or carved wood) to its inner surface (Fig. 4.8). Alternatively, a cable-tie (e.g. RS 543–349) can be

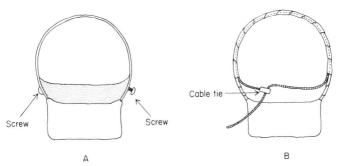

Fig. 4.8. Collars with closed loops of thick brass strip can be adjusted to the correct circumference for each animal by inserting shaped gussets of wood or plastic (*A*), or by tightening a cable-tie (*B*).

riveted to the sides of the collar, and tightened round the animal's neck (Chapter 5). The cable-tie fastening is easier to make and quicker to fasten than the shaped gusset.

The smallest open-loop brass strip collars may be fastened by crimping, riveting or soldering the ends together, or by folding the ends over each other (Fig. 4.9A) and covering the join with stiff, adhesive-lined heat-shrink tubing. Collars with more than 2–3 months of cell life should have their ends soldered together; otherwise the tag may transmit poorly, or fail completely, as the metal surfaces oxidize at their join. It is best to use brass nuts and bolts to join brass strips more than 4 mm wide. If the circumference is fixed, the ends are bent outwards and the nut soldered to one side, with the screw threaded to ensure correct alignment (Fig. 4.9B). If the collar must be very carefully adjusted, because the animal's head and neck circumferences are very similar, solder the head of a screw inside a short brass strip on one side of the tag and drill the long strip on the other side with holes at 4–7 mm

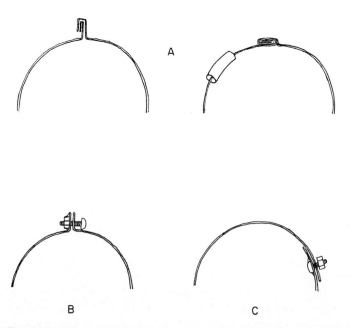

Fig. 4.9. Closed-loop collars of thin brass strip can be closed by folding the strips together and covering them with heat-shrink tubing (*A*). Collars with a fixed circumference can be closed with a bolt at the top (*B*). An adjustable collar can be closed by threading a bolt, set in one brass strip at the side, through one of a series of holes in the other strip.

intervals (Fig. 4.9C). This keeps the inward-facing head of the screw at the side, where it is least likely to rub the animal's neck, especially if its surface is smoothed with a file.

All holes for screws or rivets should be drilled in the collar strips and filed smooth before the brass is bent into shape. Hold the strips in a vice for drilling, not in the hand, or they may fasten to the bit as it penetrates them, and cut your fingers. If the brass is first punched lightly at the point to be drilled, the bit is less likely to slide off-centre. Fasten the two sides of an open loop before you assemble the tag, using the central bolt hole if the collar has an adjustable circumference. The brass strips should be covered with heat-shrink tubing, or wrapped with 5–10 mm wide strips of fabric-backed tape (e.g. RS 512–058) to protect the animal's neck. With open-loop collars in particular, it is also best to protect the antenna emergence points with silicone caulking. As with the wire loop collars, closed-loop designs must have an external means of disconnecting them for storage, whereas open-loop designs do not transmit until the collar loop is fastened.

III. CIRCUIT COMPLETION

A. Testing

Tag function and tuning should be tested before potting. The initial test, which indicates whether the transmitter is running, can be done with a microammeter. However, it is then wise to listen to the signal on a receiver, without the meter in circuit because this may substantially affect the tuning. If the pulses are very prolonged and "wheepy", you should increase the capacitance in the tuning circuitry of a home-made transmitter, either with the trimming capacitor in the transmitter or by adding a small (2.2–5.6 pF) fixed-value capacitor to the outside of the transmitter, between the antenna lead or the positive lead and the tank circuit lead at its transistor/crystal junction. If the pulse is only slightly wheepy, the effective capacitance increase on potting the tag, which is equivalent to perhaps 1 pF more in the tank circuit, will probably correct it. If the signal is more of a "click" than a "peep", the tuning capacitance is too high. This not only reduces signal strength on the correct harmonic, but can also lead to overloading of the transistor and failure of transmitters with lithium/copper-oxide cells (see below). If the transmitter has no trimming capacitor, its tank circuit will have to be removed from the potting and replaced.

This testing will not usually reveal the presence of dry joints, in which the solder has not bonded to both metal surfaces. Dry joints are one of the most common causes of tag failure. It is particularly easy to make them when soldering transmitter leads to brass strip antennas or other large areas of metal, on which the solder may cool too quickly for bonding. These surfaces should be thoroughly tinned in advance, and you must be sure that the solder on them melts when the joint is made — holding them in pliers if necessary to protect your fingers!

B. Lithium/copper-oxide cells

These cells are nominally 1.5 V, but deliver at least 2 V when first connected. In a single-stage tag, the voltage falls to about 1.8 V during the first day, and reaches 1.5 V after two or three days. If a transmitter with a ZTX 314 transistor has too high a tank-circuit capacitance, a feedback effect is liable to develop when the lithium/copper-oxide cell is first connected, such that the transmitter produces a continuous whine rather than a pulsed signal. If this condition is allowed to continue for more than a second or two, the transistor will be damaged: the tag will either fail there and then, or stop transmitting as the cell voltage drops towards 1.5 V.

It is therefore important to listen to the signal from these tags, not only when they are first connected without an ammeter in series, but also after potting, with the tag lying on an insulated surface rather than held in the hand. If a feedback effect only appears after potting, the tag can be run with a 200–500 Ω resistance in series for a day or two, by which time the cell voltage will have dropped enough for safe running with the extra resistance removed. However, the cell voltage will increase again if the tag is switched off and stored, so you may need the series resistor again when the tag is restarted.

The MMT and FMMT series transistors are less easily overloaded, and rarely suffer from this feedback effect. Nevertheless, it is wise to test all tags which contain lithium/copper-oxide cells by running them for two or three days prior to use, to be sure that they still function when the cell voltage has fallen to its nominal value.

C. Zinc–air cells

If tags powered by zinc–air cells are to be used within 2–3 months of construction, they can be potted with the sealed cell in circuit. The adhesive strip over the airhole should be peeled back for about 15 s just before potting: the oxygen which enters will be sufficient to power the tag for 10–100 min (depending on cell size), so that tag tuning can be checked before and after potting. The potting should slightly overlap the edges of the tape, and a layer of varnish (e.g. nail lacquer) can be used to ensure an airtight seal. As the oxygen partial pressure falls within the cell's air chamber, the tag's signal pulses should slow and stop: check that they do! This short period of activity seems not to affect subsequent tag life adversely, and the tags are easy to restart in the field, by scraping off the sealing tape.

D. Switches

Apart from tags which are for immediate use, and those which have zinc–air cells, solar panels or open tuned-loop antennas, there must be some way of turning off the transmitter for storage. One switching technique is to join the negative rail to the base lead of the oscillator transistor, with a wire which loops out of the potting. The loop is severed for testing the tag, rejoined for storage, and re-opened or cut off completely to start the tag in the field. Since tags stored with this sort of switch continue to draw current at 2–4% of the working load, many people prefer to expose a loop in one of the cell leads as a switch. The loop is cut for storage, leaving a pair of wires which are rejoined by soldering before the tag is used. These joined wires, and the cut wire ends of base–earth loops, should then be covered with a layer of potting or other sealant to prevent corrosion working along them into the tag.

For animals which may chew or otherwise damage tags, the switch wires should come to the surface where they are least likely to be affected. On a closed loop mammal collar, for instance, the wires are best exposed inside the loop. The same principle applies to other leads and circuit components which pass near the potting surface. Moreover, leads to brass antenna strips should be joined to the inside of the strip rather than cross its surface. On mammal collars, the potting tends to be chewed most at the sides, and to a lesser extent underneath, so vulnerable leads and components should be kept away from these areas if possible.

Sometimes it is impractical to pass wires through the tag's surface for switching, because there is too much risk of damage or because the transmitter and cell are completed enclosed in a moulded casing. A magnet-operated reed switch may then be used to complete the circuit, provided that the tag is large enough. A normally-closed (NC) or changeover (CO) switch is connected so that the tag is running, and embedded near the surface of the tag. For storage, the tag can be turned off by taping a magnet to its outer surface, along the length of the switch; check with a receiver that the magnet is strong enough and in the right place. Small CO switches, 3.3 mm in diameter and 18 mm long with their leads cut short (FEC TRC 200) can be embedded 3–4 mm below the tag's surface and turned off by a 25 × 6.5 × 6.5 mm (1 in × 0.25 in × 0.25 in) magnet (FEC RSH 34). Heat-shunt the leads if you solder to them close to the glass body of the switch, or else you may crack the glass. If you need small tags with neither external switch wires nor the added weight of reed switches, you will have to pot the completed circuit just before you use it.

E. Potting

Epoxy resins (two-component glues) make good potting compounds for many types of tag, being easy to apply and compatible with electronic components. They vary in vicosity when mixed, in setting time, and in hardness when set. Since there are many types on the market, it may be necessary to test several locally available brands to find the most suitable for a particular tag design. Rapid Araldite is particularly good, because the freshly mixed glue and hardener are sufficiently viscous to build a 2–3 mm layer in one coating, especially in fairly cool conditions. If the two adhesive tubes are warmed somewhat before mixing, the viscosity is reduced and a very fine layer can be achieved; the setting time is also reduced by the warming. Alternatively, you can obtain a fine, hard and very clear layer from RS epoxy adhesive (554–850) without any warming. When Rapid Araldite sets, the resulting hardness depends on the proportions of adhesive and hardener in the initial mix. The hardest finish is obtained with a slight (5–10%) excess of glue to hardener. The potting is then very hard, but brittle enough to crack if you drop a large tag. Mixed with a slight excess of hardener, the Araldite sets more like a hard rubber. Some epoxy compounds never become harder than this: such a set is convenient if the tags are unlikely to get damaged and you may recover them for cell replacement. For long-life tags, however, it is best to avoid "soft" mixes, because these tend to become even more rubbery over long periods in the field.

Large blocks of epoxy compounds can be drilled or sawed, but wire clippers (e.g. RS 544–386) are convenient tools for removing potting from small tags, for repair or cell replacement. These tools can be used for a variety of tasks, depending on their stage of wear. At first they should be kept for the finest copper wires. After a time, their edges become worn (usually because they have not been used *only* for fine wires!), and they are best set aside for cutting thicker wires, and then perhaps for cutting stainless steel antennas (which really ruins the edges). At this point they cannot be further spoiled by cutting Araldite. As a final ignominy, they may be allocated to field work: they are rather useful for mending wire-mesh traps or for toe-clipping.

Dental acrylic gives an even tougher coating, and is often used on commercial tags. Several techniques have been developed for coating tags with acrylic, which is supplied as a liquid with a powder catalyst. The simplest method, suitable for tags up to 8 g, is to work a thin layer of the liquid over the package, dust with powder and leave it to set. One layer is enough on 1–2 g tags, but larger tags require several applications. A second technique is to mix a small amount of powder with the liquid acrylic in advance, and then to spread the resulting mix over the tag like an epoxy, finally dusting with more powder to speed the setting reaction. Dental acrylic can also be mixed in a PTFE (Teflon) coated container and poured into moulds, but is less suitable than epoxy for this purpose because it generates much heat during setting: several pours may be necessary to avoid damaging the cells. It is also harder than epoxies to remove for making repairs or replacing exhausted cells. It does dissolve slowly in acetone, but care is needed not to dissolve the insulation from the components as well.

Dental acrylic is good for providing light, strong coatings on delicate components, but many people now prefer Araldite or other epoxies, which are simpler to apply to large tags, easier to remove, and less prone to become contaminated and set in storage containers. Araldite seems adequate for many animals that are liable to chew their tags, but in some cases the greater hardness of dental acrylic may be crucial, especially for mammals in watery environments. Water will very gradually penetrate epoxy resins, especially if they contain many air bubbles in which water vapour can collect.

However, the most common route for water entry, whatever the potting compound, is along antennas, either by surface corrosion of bare wires or by capillarity between a plastic coating and the potting. If the antenna lead and antenna are of different metals, electrochemical potentials help to drive the corrosion process, and the antenna may become electrically disconnected even though it remains in place. It is therefore essential to caulk these emergence points if tags are to last long in damp conditions. Acetic-based silicone sealant bonds well to dry Araldite and some plastics, but virtually nothing bonds to PTFE-coated wires.

In the case of implants and ingested tags, it is important not only that body fluids do not penetrate the tag, but also that the tag itself is not toxic to the animal. Epoxy resins are likely to produce toxic or inflammatory effects, and to be penetrated by body fluids. These tags should thefore be given an outer coat of beeswax or a physiologically compatible silicone compound.

A variety of other potting compounds are available. For instance, soft two-part silicone rubber (*not* acidic caulking) may be used as a shock-absorbent filler in tags with a hard outer casing, and 3M Scotchcast is often used as a pourable potting for moulded tags. Other potting compounds should be used with care, because they may contain substances which react with component coatings, or allow the migration of charge-carrying ions.

Potting in moulds is sometimes an attractive technique for mass-producing tags. The potting compound can be mixed in a plastic bag, and squeezed out through a cut corner into plastic, silicone or epoxy moulds, with access holes left for changing cells, inserting antennas or tuning the transmitter. An even simpler system for large-mammal tags is to pot the transmitter and cell(s) in a loop formed between two strips of collar material (Fig. 4.10). The bottom of the mould (the back or front of the finished tag) is formed by a plastic surface against which the strips are held. If a whip antenna is used, this passes up between the strips on the longer side of the collar.

A disadvantage with moulded tags is that they usually require more potting, to ensure that the cell, transmitter and leads are adequately covered, than if the potting is spread on the tag as a thin layer. The extra weight of potting militates against using moulds for small mammals. Even for large mammals there is a lighter alternative, in which a thin, but very strong fibre-reinforced epoxy casing forms a tight fit for the cell and transmitter (Section IV.G).

Since layered potting is best completed in one go, it is quite important for these tags to have their cell, transmitter and antenna held fairly rigidly together before the coating is applied. Cyanoacrylate glue ("Superglue") can be used to stick parts of the smallest tags together. For larger tags it is

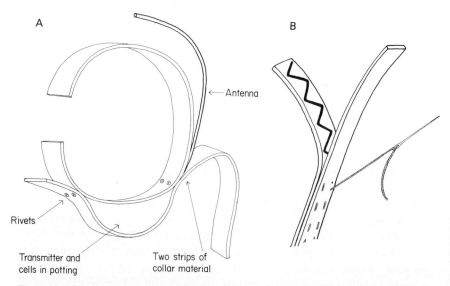

Fig. 4.10. Tags for large mammals can be potted between two strips of collar material (*A*). The antenna can be free-standing, or fixed to the antenna with heat-shrink tubing, or sewn between two strips of the collar material (*B*).

best to use small amounts of potting, which also fill up air-spaces so that the final potting can be completed well within the compound's setting time. This preparation is especially important if sensors, such as mercury switches, are to be set in a particular position.

F. Labels

Before potting tags, transfer the transmitter's frequency label to the antenna wire, or write the frequency on the cell where it will be seen through the potting. This saves a great deal of time searching for signals later on. If tags are large enough, it may also be worth incorporating an address or "reward" label, especially if you are studying game species. A lot of time can be wasted searching for signals from shot animals!

IV. TAG DESIGNS

Considering the variety of power supplies, antennas and mounting techniques that are available, a vast variety of different tags can be built. Now that the general principles of tag construction have been described, it is worth covering in more detail the making and mounting of nine different designs, which can be scaled up or down to suit a very wide range of species.

A. A 1 g glue-on

The following description is for a 1.3–1.5 g tag, based on a 650 mg transmitter with a 500 mg A312 zinc–air cell. The signal life is 3–5 weeks. The tag would weigh about 1.0 g with a 300 mg silver-oxide cell, but would then transmit for only 10 days (using a 100 μA transmitter). You will probably need to buy the transmitter itself, because it takes a great deal of practice to produce a potted circuit weighing 650 mg. However, the same tag design, with a 1 g transmitter, a 0.30–0.35 mm (0.012–0.014 in) antenna, and a 13-size or 675-size cell will make a tag of 2.5 g or 4 g, respectively, for larger animals. These tags are suitable for gluing directly to fur or feathers. They may also be attached with tape, tail-clips or harnesses to small birds, reptiles and amphibia, although zinc–air cells should be replaced by mercury or silver-oxide equivalents in tags for dirty or wet environments.

With a zinc–air cell, the seal must be uppermost on the finished tag. Since the cathode then faces downwards, the negative lead passes close to the positive cell casing. To reduce the risk of a short circuit, it is best make this lead from fine insulated wire, soldered at one end to the cathode and at the other to the negative transmitter lead near its emergence from the transmitter potting (Fig. 4.11A). On the other hand, a mercury or silver oxide cell can be mounted with the cathode upwards, and a "switch" loop formed in the negative lead before this is soldered to the cell.

Cut the chosen length of antenna from 0.15–0.25 mm (0.006–0.010 in) guitar wire, and bend 2–3 mm at one end back to form a "U" which will hook round the antenna lead. Solder the two wires, and trim the lead. If the transmitter was not pre-tuned, give it a final adjustment now, while the cell is still active.

Make a low-viscosity mix of epoxy adhesive, in the case of Rapid Araldite by pre-warming to 30–40°C. Holding the thickest cell lead in snipe-nosed pliers (Fig. 4.11B), use a screwdriver with a 2 mm blade to smear a thin epoxy layer over the cell and transmitter, also working potting into the space between them. Be careful to cover the edges of the sealing-tape on the cell, but to leave most of the tape uncovered for subsequent removal. The potting should be at its thickest where the transmitter is joined to the antenna. To complete the potting without getting epoxy on the pliers, move these to grip

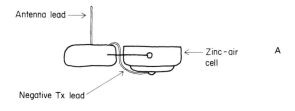

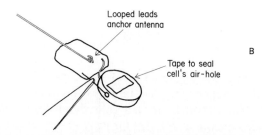

Fig. 4.11. Making a 1.5 g tag for glue-mounting. The transmitter leads are joined directly to the cell (with flux-cooled soldering — see Fig. 4.1), and the antenna held in a loop of the antenna lead for soldering. The tag can be held by the positive lead for potting.

the antenna near its base, but be careful not to bend the antenna attachment. If a mercury or silver-oxide cell is being used, hold the protruding wire "switch" loop instead of the antenna. An artery clamp can be used to hold this loop or the antenna while the potting hardens.

B. A 12 g backpack

Based on a 1.0–1.5 g transmitter and a 7.5 g LC02 lithium/copper-oxide cell, this tag will run for 9–18 months, depending on the transmitter current. The design is used mainly for harness-mounting on birds which weigh at least 250 g, although it can also be modified for wiring to reptile carapaces, for gluing to hedgehog spines and for many other applications.

Tin the cell leads near their junction with the cell, then bend them outwards and shape them to hold the harness tubes (Fig. 4.12A). Be careful not to pull the tags off the cell as you bend them, because the spot welds which attach them are not very strong. Solder 2 cm of 0.3–0.4 mm diameter

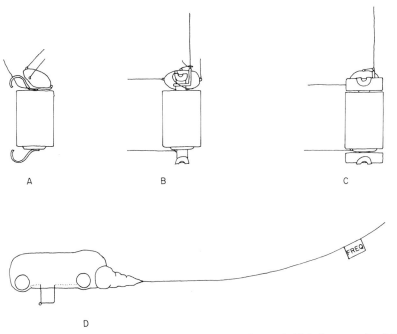

Fig. 4.12. Making a 12 g bird backpack. The cell leads are bent to hold the harness tubes (A), before the transmitter and antenna are soldered in place (B) and the harness tubes inserted (C). Before potting, make a "switch" loop in the positive lead, and attach a frequency label to the antenna (D).

copper wire (e.g. an RS resistor lead offcut) to the positive terminal, and solder the negative transmitter lead to the negative terminal so that the transmitter is held tight against that end of the cell. Make sure that the potting round the transmitter crystal is thick enough to prevent any possibility of a short circuit against the negative terminal or lead. Cut 2 cm lengths of harness tubing; thin PVC tubing with a 4 mm diameter (e.g. RS 399–271) is very suitable. These are inserted after the cell leads are attached, because the heat of soldering would deform them.

The antenna must be very thoroughly anchored in these tags. Stiff stainless steel antennas, such as 0.35–0.55 mm guitar wire, should be prepared with the inner 5 mm bent at a right angle: insert the bent end between the transmitter and cell, or under the negative cell lead. Since the antenna is positive, it must be well insulated from the negative terminal, by a covering of heat-shrink tubing or plastic with a high melting point (e.g. PTFE, *not* PVC). The antenna lead is bent to one side, to wrap round the antenna where it passes the transmitter. After the lead has been soldered, cover the antenna with a 30–50 mm length of tubing, distal to the join, to protect the antenna from acidic silicone sealant after potting. If the antenna is a flexible material, such as fishing trace covered by layers of heat-shrink tubing, the tubing must pass alongside the transmitter as far as the cell, with the antenna bent back in a dog-leg to join the transmitter lead. There is a high risk of antenna breakage at the inner end of the tubing if it is poorly anchored. Antennas can also be anchored by running them alongside the cell next to the harness tubes, to leave the tag at the opposite end to the transmitter, but in this case they must be thickly coated to minimize detuning capacitance between the antenna wire and the cell.

When you first make tags of this size, it is a good idea to glue the transmitter, antenna and harness tubes in place before testing the tag and fine-tuning it. A microammeter between the positive lead from the cell and the transmitter will indicate whether the tag is working. Then bend the transmitter lead to run along the cell, turning it outwards near the positive terminal to form a "switch" loop with the lead already soldered to the cell (Fig. 4.12*B*). However, to avoid any risk of overloading the transmitter with the initial high cell voltage, you should not solder these leads together without listening to the signal (see Section III.B). With tags of this size, fine-tuning can be done most cheaply, and with negligible weight increase, by adding a 2.2–5.6 pF capacitor between the external leads from the transmitter, rather than by using a built-in trimming capacitor.

Use a fairly viscous epoxy mix to pot these tags, with slightly less hardener than adhesive. Hold the positive end of the cell and work the epoxy into any spaces between the transmitter, antenna, harness tube and cell at the other end, spreading a generous layer over the antenna and a 1 mm layer over the other parts. Cover as much of the cell as possible before transferring your grip to the antenna and harness tube at the transmitter end. It may help to insert a rod or screwdriver through this tube while the potting layer is completed at the other end: the tag can be left suspended by this tube while the epoxy hardens. If the surplus harness tubing is trimmed off within about 6 hours of potting, using a scalpel or other sharp blade, the epoxy will cut rather than crack, and leave a neat finish. A silicone cone spread up the antenna within this period also seems to bond well to the epoxy (Fig. 4.12C). The tag can then be left for 48 hours, to check that it still runs after the cell voltage has dropped, before it is disconnected for storage.

This tag design, and the two that follow, can also be made with two-stage transmitters, using 1/2 AA lithium cells which give 2.7–3.4 V. The life is then 4–8 months, depending on the circuit current. Lives of nearly two years are possible with low-power two-stage tags and 17 g AA cells, the tag weight then being 30–40 g.

C. A 14 g posture-sensing tail-mount

This tag's 1.0–1.5 g transmitter contains a pair of parallel capacitors, which are joined by a mercury tilt switch (e.g. Gentech HG520L4), to control the pulse rate (Chapter 2.V.E). An LCO2 cell is used, giving a life similar to that of the previous tag. The light and flexible antenna is attached along the shaft of one tail feather, so the tag is best for birds with relatively long tails. Weighing about 14 g with a 1–2 mm layer of potting, this tag has proved ideal for medium-sized raptors. The weight can be reduced by about 2 g without the mercury switch, but a version with a smaller cell would be needed for birds weighing less than about 600 g (the tag is then 2.5% of the bodyweight). Tail-clips can be used on these tags, with free-standing antennas, for birds with short tails.

To make the tag, the cell tabs are folded back on themselves, so that they do not project beyond the edge of the cell. Leave the folds slightly open at first, so that they can be used to hold the mounting ties after the transmitter leads have been soldered in place. Tin both tabs, and solder the transmitter negative lead as in the previous tag. A 20 cm length of antenna wire is soldered with the 2 cm cell lead to the positive terminal. This free-standing "ground-plane" antenna, and the 30 cm main antenna, may be made of 20–27 kg nylon-coated multistrand fishing trace. However, fine 49-strand aircraft cable, coated with a tough yet flexible heat-shrink polymer (e.g. Rayfast RNF 3000), is more durable as a main antenna: fishing trace usually breaks at the distal end of the supporting feather after 4–6 months, thereby reducing the reception range. Before attaching the main antenna to the

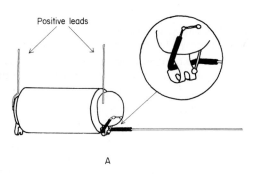

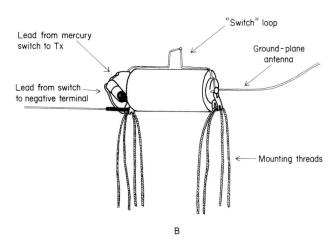

Fig. 4.13. Making a posture-sensing tag for tail-mounting. The cell leads are bent to hold the mounting threads, and the transmitter attached with its main antenna anchored behind the negative lead (*A*). The mounting threads may melt if heated, and are therefore inserted after all soldering to the cell leads is completed (*B*).

transmitter lead, bend it so that it will emerge from the potting on a level with the bottom of the tag and slightly off-centre (Fig. 4.13*A*).

If you attach the transmitter to the end of the cell at one side, there is more space for the mercury switch. Use a lentil-sized blob of epoxy to glue the switch and transmitter in place against the cell, but be careful not to cover the negative cell tab. The switch leads should point away from the cell and the main antenna, with the switch angled so that it closes at about 40° from the horizontal. When the adhesive has hardened, one switch lead is soldered to the transmitter's free capacitor lead, and the other to the negative cell tab. You have now finished soldering to the cells, so the attachment ties can be inserted and squeezed tight in each tab's fold without risk of heat damage. Fine synthetic thread for fishing nets (e.g. 30/9 Perlon) or cobblers' thread makes good ties, cut into four 30 cm lengths for each tag. Two lengths are held in the middle by each tab, so that four equal lengths hang from each end of the cell (Fig. 4.13B). At this point the circuit and mercury switch may be tested, and the positive leads joined prior to final tuning, which is easiest with the slower of the two pulse rates.

Before potting, knot the eight ties loosely together. Use this knot to hold the ties out from the tag: you can then spread potting round the ties without smearing them, so that they emerge cleanly from the tag. If you start by holding the knot and positive end of the cell in one hand, the epoxy can be worked into the crevices round the transmitter and mercury switch, and the tag then suspended from the knot while the other end is potted. Finish the tag by making a silicone cone to protect the emergence of the ground-plane antenna.

D. A 14 g necklace

This is a LC02-cell design, suitable for birds of at least 400 g. For lighter species, it can be built with smaller cells. With a whip antenna made of fishing trace, this tag has proved very satisfactory on galliformes, being quicker to attach and giving better ranges than the previous tag designs. However, birds which fly frequently (e.g. passerines) have a tendency to catch the antenna under a wing, which irritates them. It may be possible to prevent the resulting behavioural disturbance by using a guitar wire antenna, which is stiffer and lighter than fishing trace, but this has yet to be tested.

The necklace cord is a 25–30 cm length of hollow, soft, braided artificial fibre, with 5–7 mm diameter unstretched. The "woolly" cording often used for football-boot laces is ideal, with lesser diameter lacing for smaller tags. Do not use cotton lacing, which wears out within a few months. Through this cord is threaded an 18 cm length of 20–27 kg nylon-coated fishing trace, which is soldered, along with a 2 cm wire lead, to the cell's positive tab. Fold the tab back on itself to grip the cording. The negative cell tab is tinned near its base and bent to hold a 2 cm length of 4 or 6 mm diameter PVC tubing (e.g. RS 399–271), through which the necklace cord will be threaded after potting. To prevent heat damage to the tubing, attach the transmitter to the tab by its negative lead before inserting the tubing.

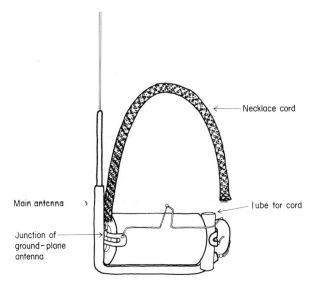

Fig. 4.14. The component layout of a 14 g necklace tag for game birds. The necklace contains a ground plane antenna, which is attached to the cell's positive terminal.

The main antenna is a 27 cm length of fishing trace, which is coated with two layers of heat-shrink tubing to stiffen and distance it from the cell. The inner length of tubing extends for 20 cm, and the outer for 15 cm. Bend the antenna through a right angle about 3 cm from the transmitter end, and solder it so that it passes along the side of the cell and up parallel with the necklace cord (Fig. 4.14), where it should be held in place by a dab of epoxy or cyanoacrylate glue. A 3–6 mm gap should be kept between the main antenna and the ground plane: the tuning will be adversely affected if the main antenna passes too close to the ground plane. Although the necklace cord should not be threaded through the tag's tube until after potting, during the final tuning it should lie against the tube, and thus bring the ground-plane antenna close to its ultimate position. Do not allow the main antenna to lie alongside the ground plane, which can result in feedback and thus destroy the transistor.

After potting, and trimming off any excess plastic tubing, the necklace cord's emergence must be sealed to prevent moisture from taking this route into the tag. Since the sealant should be worked thoroughly into the cord, which provides an access route to the connections on the positive cell terminal, you must use a silicone rubber which does not gives off acid on curing. Dow-Corning 3140 RTV (FEC 101 713) is a suitable compound.

E. A 2 g collar

There are at least three ways of making a small mammal collar from a 650 mg transmitter and a small button cell (e.g. an RM312 mercury cell). The tag can have a closed or open tuned-loop antenna round the neck, or an integral loop encompassing the transmitter and cell alone. In the first two cases the tag is slightly heavier than the third, but has 2–3 times as much range. With an RM312 cell, a signal life of 20–30 days is likely in each case. Larger tags can be made along similar lines, with RM13, RM675 and RM625 cells, and will transmit for up to about 6 months (if they survive damage for that long).

Solder the positive lead from the transmitter to the cell casing so that the lead is bent halfway round the cell, and make a 2 mm diameter loop in the free end of the lead. Although the negative transmitter lead can be joined straight to the cathode, with a switch loop if required, circuit testing is easier with a separate 10 mm lead on the cell, and this should now be soldered in place. Both the crystal can and the cell casing should face outward in the finished tag, to protect the more delicate components. You can make a slight angle between the transmitter and the cell to give a good fit round the neck, but be careful that the cell does not short against any transmitter components.

If the transmitter lacks a variable capacitor to tune the loop, solder one between the antenna lead and the positive lead or crystal can: the 5–25 pF Johansson/Tekelec trimmer (AT9402–6) is small enough to fit flush against the transmitter potting. Trim the antenna lead to a 2 mm diameter loop: this anchors one side of the collar, the other side being attached to the loop in the positive lead (Fig. 4.15A).

For a closed-loop collar, cut a length of really tough wire (e.g. PTFE-coated multistrand copper) about 2 cm longer than the maximum expected head diameter. Strip the coating from about 2 mm at each end, tin the wire and bend it into a hook. The hook at one end should be soldered to the antenna lead loop, taking care that the variable capacitor remains well

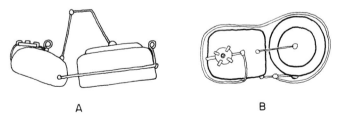

A   B

Fig. 4.15. The component layout for two small tags. The tuned-loop collar has yet to be attached to the loops on tag *A*. Tag *B* has an integral tuned loop wound round the cell and transmitter.

soldered to the lead too. Fold the collar so that it doubles back in the "sliding-8" format for about 12 mm close to its centre, and slip an 8–9 mm length of tight-fitting heat-shrink tubing over the doubled piece of wire (refer back to Fig. 4.5). Attach the other end of the collar to the wire loop on the cell.

You can use similar collar wire for an open loop antenna. Cut a length about 2 cm greater than the greatest expected neck diameter, form a small tinned hook at each end as for the closed neck loop, and cut the wire in half. Carefully adjust a wire stripper to the diameter of the conductor, and sever the insulation at half the minimum neck circumference from each hooked end. Do not strip the insulation off the wire, because it will prevent oxidation until you are ready to mount the collar, but expose about 1 mm of the conductor so that the collar sides can be joined briefly to tune the loop. Solder the hooked end on one wire to the positive lead and the other to the antenna lead. To mount the tag, slip a 15–20 mm length of narrow heat shrink tubing over one of the collar wires, and pull the surplus insulation off both of them. Twist them together, solder the join if possible, and shrink the tubing over the bare metal.

An integral tuned loop is formed from 10 cm of 0.22 mm (34 s.w.g.) enamelled copper wire. Tin about 2 mm at one end and hook it round the base of the antenna lead, then wind the other end horizontally twice round the cell and transmitter, before soldering it to the positive lead (Fig. 4.15B). If necessary, cyanoacrylate glue can be used to anchor the turns. The "unused" loops on the antenna and cell leads can be used to anchor the collar material, but must not be connected electrically: if a wire is to be used round the neck, one end should be soldered to the crystal can rather than to the antenna lead.

After testing the circuit, hold the tag by the antenna wire, or with an artery clamp on the negative lead, while you pot it. Use a 10% excess of adhesive in the epoxy mix, and make sure that the collar joins are well potted, but do not cover the tuning lug of the variable capacitor. This lug should be tweaked, to free the rotor, after the potting has dried for 20–30 minutes. When the loop is tuned, the pulse rate will slow and the signal increase greatly in volume as it reaches resonance. A little acidic silicone sealant may then be smeared over the lug, and around any antenna emergence points. Collars which are stored with the loop open can have their negative leads soldered and covered with a thin layer of epoxy, but the other types should be disconnected unless they are for immediate use.

F. A 25 g thermistor collar

This closed brass-loop collar design has given good results on grey squirrels, transmitting a temperature-modulated signal for 5–9 months from the LC02 cell. Without the thermistor, the tag will theoretically run for up to 18 months, but should not be relied on to run for more than about a year without damage.

Use a metal guillotine to cut a smooth-edged strip of 0.8–1.0 mm brass about 9 mm wide. Cut a 170 mm length of this strip, and drill a 3 mm (⅛ in) hole about 45 mm from each end. Bend the strip to shape, and tin it where the cell, antenna and tuning capacitor leads will join (Fig. 4.16). Fold the cell's positive tab back on itself in line with the edge of the cell, with enough space for it to slip over the brass strip. Fold the negative tab along the side of the cell, and tin the outer face of both tabs.

Solder the transmitter's antenna lead to the inside of the brass strip so that the lead remains covered when the transmitter is pushed back against the strip, with its positive lead pointing upwards and negative lead to one side. Make sure that neither of these other leads, nor the thermistor lead, is within 1 mm of the brass. Check that the gap between the transmitter and the other side of the collar is not greater than the length of the cell (Fig. 4.16A). Insert the cell, with its negative tab on the same side as the negative transmitter lead, and its folded positive lead gripping the brass strip, to which it should be soldered. The positive transmitter lead is then bent along the top of the cell and soldered to the positive cell tab on the inside of the brass strip. Make sure that neither the crystal nor the transmitter leads come within 1 mm of the negative terminal or tab: insert a strip of insulating tape if there is any risk of a short circuit. If a 2–10 pF variable tuning-capacitor is to be used (e.g. Johansson/Tekelec AT9402–6), this can now be soldered across the bottom of the loop.

In a squirrel collar for monitoring drey use, the thermistor is positioned between the ends of the brass strip on the tag's bottom surface. The tip of the thermistor is thus exposed to the air during foraging, but should not stand

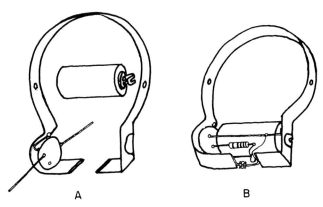

Fig. 4.16. Making a squirrel collar. The cell and transmitter leads are bent to shape before they are soldered to the brass antenna loop (*A*). The thermistor and loop-tuning capacitor are added (*B*) just before potting.

proud of the potting. A miniature glass bulb type, with a resistance of 470 kΩ at 25°C (e.g. RS 151–164), is ideal as a direct load thermistor (Chapter 2.V.E). If the transmitter has a 1 MΩ load resistor to parallel the thermistor, a 22–330 kΩ resistor should be soldered in a series with the thermistor when it is connected to the appropriate transmitter lead. The other thermistor lead may be soldered to the brass collar strip under the cell, where the tuning capacitor is connected, provided that this is done on the same side as the positive cell connection (Fig. 4.16*B*).

After testing the circuit with a microammeter, the negative transmitter lead is soldered to the nearby cell tab, with a "switch" loop if the tag is not for immediate use. Strengthen the tag by winding three or four turns of braided nylon or terylene thread round the slight neck at the base of the loop, to reinforce the potting. A viscous epoxy mix is needed to give a good coating, with a 10% excess of adhesive in the case of Rapid Araldite. Work the potting well into all the spaces between the cell, transmitter and brass strip, covering all parts except the thermistor tip with a layer 1.5–2 mm thick, but leave the tinned ends of the brass strip exposed if a tuning capacitor has not yet been chosen. If you are using a trimming capacitor, this should be surrounded but not covered by potting, and should be checked for free rotation within 30 min. When the potting has set for several hours, the tuning may be set with the trimmer, or by soldering and removing different fixed value plate ceramic capacitors until the signal is strongest. The capacitor should then be covered with fresh potting.

The cable-tie fastener (e.g. RS 543–349) is fixed in place with 3 × 10 mm (⅛ in × ⅜ in) pop-rivets, which are inserted from the outside of the collar. Two holes must first be drilled along the centre-line of the 4.7 mm wide tie, with the first about 15 mm from the head end. Finish the tag by binding all the exposed brass with adhesive fabric-backed tape (e.g. RS 512–058), using a double layer over the rivets.

This collar can also be built as a fixed-diameter open loop, joined by a screw at the top (Fig. 4.9*B*), and is then suitable for rabbits or other mammals weighing more than about 500 g. The general format, in which the brass antenna material is folded round the components to give strength and protection, can be used with smaller cells and thinner brass to build a wide range of collars with fixed and variable diameters. With ingenuity, the width of the lower part can be reduced to little more than the length of the cell, by building the transmitter flat along the upper surface of the cell or even around its ends.

G. A 125 g collar package

As mentioned in Section III.E, there are many ways of making moulded or cased packages for large animals, mainly using two-stage or even more powerful transmitters. If you make your own moulding, be careful that the transmitter and cell are supported away from the side of the mould, and ensure that any attachment lugs or antenna emergence points are strong enough to withstand severe treatment. It is better to make collar packages wide or thick rather than deep, to minimize any pendulum effect.

The simplest approach is to buy a commercial casing (e.g. Biotrack C-casing) and add your own transmitter, antenna, cell and switch or sensor system. The casing is supplied with a pair of 4 mm holes for bolts or rivets in each attachment lug, but the exit channel for a free-standing whip antenna is not drilled, to allow the casing to be used with an integral loop antenna or with a whip attached to the collar material. A free-standing whip antenna can be made from 3 mm steel brake cable, covered with one or two layers of tough, heat-shrink tubing with a meltable lining (e.g. Raychem SCL). Drill the antenna channel first with a 2 mm bit, and then enlarge it with a bit of the same diameter as the coated antenna, which should fit quite tightly. Smear the heat shrink close to the tinned end of the antenna with silicone sealant before fitting it, to lubricate and seal the insertion. Any risk of the antenna being pulled out must be eliminated by soldering the tinned end to a piece of brass shim, which should fit tightly against the inside of the casing. Solder a 3–5 cm antenna lead to this brass or to the antenna itself. If the whip antenna will be fastened to the collar material, use bicycle gear cable coated with heat-shrink tubing, and solder on a lead before gluing that end of the antenna down one inside corner of the casing. If you need an integral tuned loop, use a 6–8 mm strip of 0.8–1.0 mm brass, or picture wire, fitted horizontally round the inside of the casing (after inserting the cell and transmitter).

If you want a tilt-switch for activity or mortality sensing, this should now be fixed with epoxy adhesive to the bottom of the casing at the antenna end. The cell goes in the casing at the opposite end, with its positive lead nearest to the antenna, and the transmitter fits in the remaining space at the antenna end (Fig. 4.17). If a magnetic reed-switch is required, this is soldered to the negative lead, and fits next to the casing beneath the cell. Tags without a whip antenna should have their integral loop added at this stage. The leads can then be joined and a provisional setting found for the tuning.

For extreme strength, the casing can be filled entirely with a fairly fluid epoxy resin, such as slow-setting Araldite. A tuning tool can be left inserted as the epoxy is poured, so that the final adjustment can be made and the tool withdrawn before the potting sets. However, it is extremely difficult to replace cells or effect repairs in tags filled with epoxy. The alternative is to fill the casing up to the top of the cell with a softer compound, such as two-part silicone rubber (e.g. Silastic 9161 RTV as FEC 101–778). A non-corrosive solidifying foam could also be used to make a very light package. It is best if the tuning can still be adjusted after these compounds have set, so that final

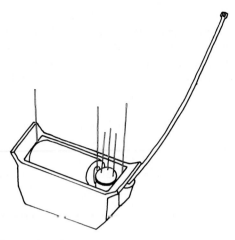

Fig. 4.17. Cell and transmitter in place within the casing of a large-mammal tag, prior to joining leads and potting.

tuning can be completed when the casing is topped-up with an outer 2–4 mm layer of pourable epoxy. This hard outer layer provides a strong "sixth face" to the tag, but can be fairly easily drilled and cracked away if you need to get at the cell and transmitter again.

Be careful not to apply great pressure to the attachment lugs when these are bolted or riveted to the collar material. It is wise to spread the load, by using washers or a drilled brass strip between the lug and the nut or bolt. A metal strip on the inside of the collar will prevent rivets eventually pulling through a soft collar material. Be careful in your choice of collar material, and make trials on captive animals if you can. Soft plastic or machine belting is sometimes used, but these materials tend sooner or later to crack. Braided nylon strapping can be used, but this sometimes frays and may then attract snags. Good quality leather is often satisfactory: it should be chrome-tanned for long-life collars. Remember that collar material may be quite heavy: a completed package can weigh 200 g, nearly twice as much as the radio components and casing alone. With 150–300 μA transmitters, the 5 A h lithium C cells will power these tags for about 2–4 years.

H. A 2.5 g implant

Based on a 13-size mercury cell and a transmitter weighing no more than 1 g, this tag will run for 4–8 weeks as a location aid, or somewhat less if used with a direct-load thermistor for temperature sensing. With a 600 mg transmitter and a smaller cell, a total tag weight of about 1 g is possible. The design can also be scaled up, in which case most field projects will favour using two-stage transmitters, to increase the range of the relatively inefficient implanted or ingested tags, rather than merely using larger cells to extend the life of a single-stage tag.

Solder the transmitter's positive lead to the side of the cell casing, with the negative lead formed into a switch loop to the cathode. The antenna is about 7 cm of 0.22 mm (34 s.w.g.) enamelled copper wire, which is soldered at one end to the base of the antenna lead and wound twice round the transmitter and cell before being joined to the positive lead. If the transmitter has not been bought with a tuning capacitor, a 5–25 pF subminiature trimmer should be soldered between the antenna and positive leads. The construction is thus very similar to the third version of the small mammal collar (Fig. 4.15B), without the collar attachment.

If a less compact tag can be used, you will get a stronger signal by increasing the size of the antenna loop to make a padlock-shaped tag. Moreover, increasing the thickness of the antenna wire increases the radiation efficiency. As loop size increases, you will reach a point at which the signal is strongest from a single turn.

A number of coating agents have been used to seal the potted tags from the action of body fluids, including layers of silicone, beeswax, dental wax, or acrylic. Seek the help of a physiologist in deciding what will be best for your particular application.

I. A 3 g tag for fish

A radio tag developed at the Ministry of Agriculture, Fisheries and Food Fisheries Laboratory in Lowestoft (commercially available from D.W. Mackay), is suitable for ingestion, implantation or external tagging of fish (Solomon and Storeton-West, 1983). The smallest version is based on 8 mm wide transmitters with D932 or D933 silver oxide cells in Sarstedt 55.480 polystyrene test tubes, which have an external diameter of only 9.7 mm. These tags are suitable for external mounting on 300–500 g fish, running for 3–6 weeks with a 75 mA h D933 cell and a medium power single-stage transmitter, but much larger tags can be built along the same lines. For instance, a 16.5 mm diameter tube (Sarstedt 55.481) could be used to build a 10-month tag, containing a single-stage transmitter and a 1400 mA h LC02 cell, or a 3–4 month two-stage tag with a 3 V lithium cell of the same size.

At least two test tubes are needed for each tag, the blunt end being cut from one to seal the open end of the other, which is cut a little shorter than the length of the transmitter and power supply. Unless the tag is for immediate use, there should also be room for a magnet-operated reed switch. More than two tubes are needed per tag because cutting often fractures the plastic, even if a fine hacksaw is used. The successfully cut ends have their surfaces ground to fit, and are joined with polystyrene modelling cement, which is also used to attach a narrow longitudinal keel with mounting holes to the surface of external tags. The external tags radiate most efficiently with a whip antenna, which can be of fine PTFE-coated wire for the smallest tags and perhaps fine fishing trace for larger versions. However, there is then a risk of water penetration where the antenna emerges. Although tags with one-month life have been sealed internally with Araldite at this point, epoxy adhesives do not bond very well to polystyrene or PTFE. The tags should be checked thoroughly for leaks before use in the field.

5

Tag Attachment

I. AVOIDING ADVERSE EFFECTS

Zoologists who tag animals have a moral as well as a practical obligation to ensure that there is no adverse affect on their subjects. In the long term, the moral aspect tends to override all other considerations, because tags which are implicated in the injury or death of their wearers are liable to become restricted or banned by law. The need to use tags cautiously is reinforced by the current widespread concern for animal welfare.

The adverse effects of tagging can be divided into long-term and short-term categories (Tester, 1971), and in each case they may be chronic or acute. Most effects are chronic and of short duration, such as a temporary reduction in foraging activity or an increase in preening or grooming at the expense of other behaviours (Boag, 1972; Bohus, 1974; Wooley and Owen, 1978; Nenno and Healy, 1979; Leuze, 1980; Sayre *et al.*, 1981; Birks and Linn, 1982). Such effects are probably present after most tagging, even if they are not recorded, and may be as much a latent result of the capture and handling of the animal as a result of the tag's continuing presence. Chronic short-term effects may delay data recording for a few days, without being seriously detrimental to the animal.

However, short-term effects can sometimes present acute problems for the animal. Hares, for example, seem especially prone to capture-shock (Keith *et al.*, 1968), which may be exacerbated by the subsequent presence of a tag, leading to death from chilling or predation (Mech, 1967). Other animals which normally adjust well to tagging may not do so at critical times of year: the disturbance alone may be enough to cause desertion by incubating birds (Amlaner *et al.*, 1978), or for females to abandon young broods (Horton and Causey, 1984), or to stop a dispersing animal securing a territory, or to tip an animal over the brink of starvation when food is short. Female ungulates have sometimes rejected their tagged young (Beale and Smith, 1973; Goldberg and Haas, 1978; but cf. Coah *et al.*, 1971). Zoologists should be aware of these possibilities, and plan the tagging operations accordingly.

Chronic long-term effects can result from badly fitting tags, which chafe or cut the skin, or cause feather-loss which reduces a bird's insulation and thus increases its energy expenditure (Greenwood and Sargeant, 1973). Overloading may reduce a bird's flying efficiency, or an arboreal mammal's foraging agility. Such effects may be enough to decrease survival (Gilmer *et al.*, 1974; Perry, 1981), if only for low weight individuals (Johnson and Berner, 1980). An increase in tag weight and complexity can decrease breeding success (Sibly and McCleery, 1980), and may also affect survival: Warner and Etter (1983) found a reduction in pheasant survival if tag weight was above 27 g. On the other hand, Hines and Zwickel (1985) found that survival increased in young blue grouse as tag weight increased from 2.1 to 6.2% of body weight, and Snyder (1985) found no tag-weight/survival relationship in another pheasant study. These authors pointed out that tag-weight/survival relationships may be artifacts of putting different tags on different groups of birds: for example, Warner and Etter tended to attach the heaviest tags just before breeding (when females are particularly vulnerable to nest predators).

Of course, some chronic effects are slight, and may not influence the recordings either directly or indirectly. For example, a harness may reduce a bird's tendency to fly (Michener and Walcott, 1966; Ramakka, 1972), and thus indirectly reduce its success in aerial courtship, but not affect its feeding behaviour on the ground. With the absolute proviso that the animal is not in pain, slight long-term effects of this type may be acceptable as a last resort.

Acute long-term effects may also be acceptable at times. Imagine that a tag makes your animal more conspicuous to a predator, or reduces its chances of escape. This may be relatively unimportant if your sample sizes are large and you are not studying anti-predator behaviour, or mortality. Even in mortality studies, you may be able to quantify the animal's disadvantage, and thereby adjust the final results.

In short, researchers should be aware that tagging is likely to affect any animal in the short term, albeit slightly, and that there may be serious effects to avoid, perhaps by modifying the tag design.

If a tag design is new, it is especially important to test it adequately, preferably on animals in enclosures before conducting a pilot study in the wild. Although many comparisons between tagged and untagged animals are difficult to make in the wild, data can often be collected on recapture rates and weight changes, both within and between trap sessions, and on reproductive status or success. Recapture rates and weight changes are the least reliable, for two reasons. Firstly, tagged animals might suffer higher mortality than controls, but have the same recapture rate because they emigrate less (H. Smith, 1980). Secondly, animals can probably compensate for foraging disadvantages by feeding for longer, and thus maintain their weight: there is growing evidence that small-animal weight trends are generally determined by internal factors, on a seasonal basis, rather than by feeding difficulties. Individuals have been found to increase in weight immediately after tagging (Kenward, 1982a), possibly because they have responded to the presence of a tag by increasing their body reserves, in the same way that small birds respond to cold weather. With conspicuous species it is relatively easy to watch feeding and other behaviours, to see whether they change after tagging.

Mounting techniques which are already known to have adversely affected some species should be treated with caution. For instance, many studies have recorded adverse affects of harnesses, on feeding, flying, moulting, breeding and survival in birds. These problems may have been due in part to inexpert harness attachment, so it is wise (and morally proper) to precede your own tagging by practising harness attachment with someone working on a similar species or, if that is impractical, to tag some captive birds. Testing for adverse effects can also give rise to ideas for design improvements to alleviate them. For example, the recapture rate with an early squirrel collar design was lower for tagged animals than for untagged controls. When a deal squirrel was examined, the cable-tie fastener was found to have rubbed the animal's neck, because the tie was riveted too far up the collar sides. Squirrels with modified collars, in which the cable ties were attached towards the bottom of the sides, did not differ from controls in recapture rates, weight changes or breeding performance (Kenward, 1982a). In a study of snowshoe hares, on the other hand, the first collars were too loose: they tended to catch in the animals' mouths until the design was changed (Brand *et al.*, 1975).

II. SEDATION DURING TAGGING

Large animals generally have to be sedated during tagging, to prevent injury to the biologist or to the animals themselves. Sometimes they are caught with anaesthetic darts, which may themselves be radio-tagged to help find the drugged animal or to locate lost darts (Lovett and Hill, 1977): the latter is especially important if etorphine (Imobilon) is being used, because of the high toxicity of this morphine derivative to humans who might stumble on a lost dart.

General anaesthesia is also necessary for implanting tags, whereas local anaesthetics, such as spray-on Novocaine, can be used during the attachment of tags to ears, wing patagia or fish fins. Mammals which are difficult to tag without getting bitten have been anaesthetized successfully with ether, or with more refined inhalants such as methoxyfluorane (Hardy and Taylor, 1980), or with intramuscular agents such as ketamine hydrochloride

(Cheeseman and Mallinson, 1980; Harris, 1980) or phencyclidine hydrochloride and acepromazine (Mech, 1974).

If possible, however, it is wise to avoid sedatives or general anaesthetics while tagging birds and small mammals, because of the risk that the animal has not recovered completely when released. The animal may appear quite normal until it exerts itself, in flight or running up a tree, and then fall or misjudge a landing, possibly with fatal consequences. Bearing in mind that humans often feel dizzy or ill for some time after being anaesthetized, despite being able to move about normally, it is quite possible that the feeding or anti-predator behaviour of other animals is adversely affected for some hours or days after they have been drugged.

III. ATTACHMENT TECHNIQUES

A. Glue-on

A rubbery epoxy mix, such as Rapid Araldite with a 10% excess of hardener, is suitable for gluing small tags to the fur of bats (Stebbings, 1982) or other small mammals. Smear the tag lightly with the adhesive and push it into the back fur over the scapulars, working the surrounding hair into the adhesive. If possible, the tag should be almost covered with hair, although the airhole of zinc–air cells must of course remain clear. The tag will detach some time between 10 days and several weeks later, depending on the species and the amount of hair stuck to the tag. Hair regrows fairly rapidly on the resulting bare patch.

If the tag is to be glued to a bird's back, stick it first to a piece of gauze, for instance by setting the freshly potted tag down on the fabric so that the epoxy penetrates the weave before hardening. The gauze should extend beyond the tag by at least one tag-width on either side and at each end. Swab the feathers in the mounting area with acetone or alcohol to degrease them, and smear some cyanoacrylate glue on the underside of the tag. Then push the feathers gently apart, position the tag so that it sticks near their bases, and smear more glue through the gauze to contact the feathers all round. Avoid putting glue on the bird's skin (or your fingers!).

A variation on this technique is to glue the fabric base to the bird before sticking the tag to the base. Raim (1978) trimmed the spinal tract feathers in the interscapular region until only about 1.5 mm of the shaft remained, and then attached the base cloth with eyelash adhesives (the Revlon and Andrea makes were best). When the adhesive had dried, a slip of cloth sewn to the transmitter was coated with quick-drying glue, and smoothed against the base fabric with a spatula to ensure a thorough join.

B. Harnesses

Many materials have been used for bird harnesses, including neophrene, silicone rubber, PTFE and other plastic tubing, copper wires covered by these materials, clothing elastic, and cording made from nylon, terylene or other braided synthetic material. Researchers differ in their choice of material and their preferred harness design. Although there is little published information to use in judging which is best, for a given species or in general, PTFE (Teflon) ribbon is at present very popular for large birds in North America (T. Dunstan, personal communication), and clothing elastic has proved satisfactory for short-term attachments in several British projects (Amlaner et al., 1978; Hirons and Owen, 1982). Monofilament nylon or fine thread should not be used, because it is liable to cut the skin at the front of the wings. The use of meshwork vests, which work their way under the feathers (Lawson et al., 1976), is a way of avoiding the straps altogether.

Although transmitter packages have been slung under the breast of owls (Nicholls and Warner, 1968) and some other species (e.g. Siegfried et al., 1977), the most popular harness format is modified from a backpack design published by Brander (1968). It makes use of a body-loop which passes behind the wings from the back of the tag, and is connected ventrally to a smaller neck-loop. This second loop is connected to the front of the tag ahead of the wings (Fig. 5.1A). The loops and tie may be formed from separate lengths of material (e.g. Dunstan, 1972), but some people prefer to thread the cord first through the posterior harness tube, tie the ends ventrally to form the body-loop, and then run them parallel along the breast to another knot which forms the bottom of the neck-loop (Fig. 5.1B). The ends may then be knotted after passing through separate tubes at the front of the tag, to secure the length of each side separately (Fig. 5.1C).

In another variation on this theme, the harness cords cross under the breast bone, where they should be tied together (Fig. 5.1D). This version is used on small birds. In some game-bird studies, the harness cords have simply passed a short distance round the base of each wing, without meeting ventrally (Fig. 5.1E). However, this harness format is liable to trap the upraised wings above the back on birds which fly frequently, and is probably best avoided.

In fact, it is well worth avoiding harnesses altogether for species which can be tagged satisfactorily in another way, because harnesses are potentially there for life, and even the best-fitting ones may eventually snag. If you must use harnesses, take great care in fitting them, and consider keeping the animals in a dimly lit enclosure for a few hours until any initial panic reaction has subsided. Game birds, for instance, often throw themselves on their

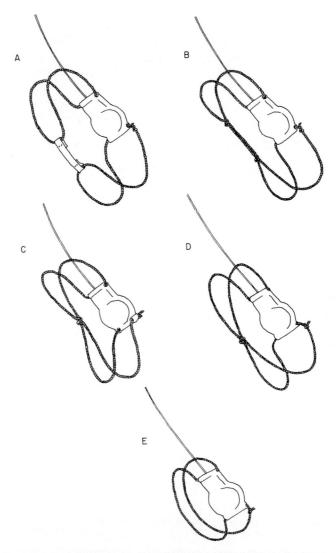

Fig. 5.1. Five ways of attaching backpack harnesses to birds (see Section III.B for explanation).

backs when first fitted with harnesses (Small and Rusch, 1985), and this must increase their susceptibility to predation if they are released immediately.

Harnesses have also been used for some lizards and amphibia, and for mammals which have narrow heads relative to their necks (Fullagar, 1967; Kruuk, 1978). Leather straps have proved satisfactory on many mammals.

5. Tag Attachment

The leather should be of a type which stays the same length when wet: split rawhide has been used on badgers (Cheeseman and Mallinson, 1980), but whole chrome tanned leather, which is supple but durable, may be better. To avoid changes in harness dimensions at sea, neophrene netting has been used on seals (Broekhuizen et al., 1980) and other plastic tubing on turtles (Ireland, 1980).

C. Tail mounted tags

There are two main types of tail-mounted tag: those which are clipped (Bray and Corner, 1972), glued (Fitzner and Fitzner, 1977), taped (Fuller and Tester, 1973) or sewn (Dunstan, 1973) to the rectrices and have free-standing antennas, and those where the tag and main antenna are glued and bound along one feather (Kenward, 1978; 1985). One of the most popular tail-clip designs has the tag mounted along a perspex plate across the top of the tail. Fine bolts are set downwards from each end of the plate, passing between the feather shafts so that a similar plastic plate can be screwed up against the lower feather surfaces (Fig. 5.2). Some glue may also be used, to anchor the tag really firmly, and to seal the nuts in place.

The number of feathers held between the plates is a compromise between the need to ensure a firm mounting, without the feathers being stressed enough to moult prematurely, and the need to avoid hindering them spreading and moulting. Ideally, the plate should fasten only the two central rectrices, which tend to moult together, but in practice it may be necessary to anchor four or six between the bolts, especially on small passerines.

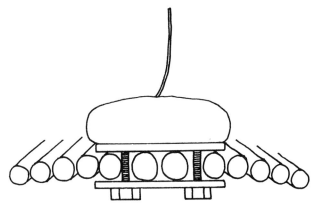

Fig. 5.2. A tail clip design. The tag is mounted on a plastic base-plate on top of the tail feathers, with another plate screwed tight against the undersurface of the feathers.

The second type of tail-mount is suitable for attaching tags to relatively lengthy feathers. To equip raptors with the tags shown in Fig. 4.13B, wrap the bird in a towel or custom-made straightjacket (Fuller, 1975), taking care to immobilize the feet. Cover the bird's head, because this discourages struggling, but leave the tail exposed. Slip a large filing card under the two feathers chosen for mounting, right up to their bases. It is easiest to use the two central rectrices, because these lie slightly above the others. However, if the tag must remain some time into the moult, mount it on the second and third feathers outwards from the central pair, with the main antenna on the second feather. The second pair outwards moult late in most raptors, unless they failed to moult the previous year. Push the upper tail coverts under the card on either side, and hold them in place there with a strip of insulating tape. Use another strip of tape to fasten the card across the feather which will not carry the main antenna (Fig. 5.3), and trim the down from the base of both rectrices with fine scissors.

Find the pair of ties nearest the tag's main antenna emergence, and thread one through a needle, ideally a small curved suture needle. Holding the mounting feather to prevent its base being stressed, pass the needle horizontally through the vane 3–4 cm from the feather's emergence, so that the tag is mounted on the feather without quite touching the skin. Fine pliers are useful for gripping the needle. Wind the anterior pair of ties on the same side two or three times round the same feather and glue them firmly to it. They

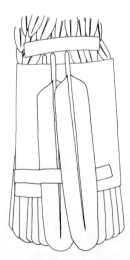

Fig. 5.3. A raptor tail ready for tag mounting. A card has been inserted under the two central rectrices, and held in place by tape across one of them. The free feather is the one to which the transmitter will be sewn. The down has been trimmed from the feather bases, and the tail coverts are held back by more tape.

must be firmly anchored, but need not be sewn in place. The ties on the other side of the tag are tied fairly loosely round the adjacent feather, and the knots sealed with glue. In this way, the tag is supported by both feathers, but they can moult independently.

Use the ends trimmed off the ties, or similar thread, to bind the main antenna to the feather vane close to the transmitter. At this point, a "soft" mix of viscous epoxy (e.g. Rapid Araldite with a 10% excess of hardener) should be used to seal the knots, to glue the tag to the main mounting feather and to fasten the main antenna for its first 1–2 cm to the vane. Make sure that there is no gap into which the bird might get its beak between the feather and the main antenna. Then use fine thread to bind the main antenna at 1–2 cm intervals along the rest of the feather vane (Plate 4), with a more flexible glue than epoxy adhesives. Evostik or Casco contact glue adheres well to the feather and to some antenna materials.

With practice, the whole mounting process can be completed in 20–25 minutes, although it may well take 40–45 minutes the first time. The bird may struggle briefly two or three times during this process. About 1% of goshawks show more severe symptoms of distress, panting very rapidly and struggling convulsively every minute or two. They die of cardiovascular collapse if the tagging process is continued. Such birds should be unwrapped without attempting to complete the mounting, and preferably released immediately.

When the tag is mounted dorsally on the tail, which is easiest for the biologist, it is also most accessible for the bird. However, the worst that goshawks have done is to remove part or all of the ground plane antennas. Tags have also been taped to the ventral surface of raptor feathers, but these could interfere with copulation. One disadvantage of all tail-mounting techniques is that some birds prematurely moult the tag. However, this problem is compensated by the lack of other adverse effects, the safe shedding of tags at the next moult, and the opportunities for behaviour telemetry.

D. Necklaces

The necklace design in Fig. 4.14 is among the easiest of all tags to mount. Once the necklace cord has been threaded through the tube at the other end of the tag, slip the loop over the bird's (or mammal's) neck, and knot the cord below the tube at the required length. The knot can be sealed with glue, but it is easier to re-use tags if you sew through it with thread, which can later be cut. Leave the free end to hang, if it is not too long, to avoid shortening the ground-plane antenna inside it and for retying if you decide to re-use the tag.

Necklaces should be long enough to let animals swallow large food items without choking. The circumference may even be slightly greater than a bird's head, provided that the neck feathers are stiff enough to prevent the tag slipping forward from an attachment position well down the neck. It is quite important to err on the generous side: enough room for swallowing an acorn was assumed to be adequate for pheasants, until a laying female tried to swallow a large snail! In contrast, a tighter fit has given no problems on Swedish black grouse, which are similar in size to pheasants but probably never eat anything much larger than birch catkins.

E. Collars

Similar considerations about swallowing apply to the attachment of collars, which need to be fairly tight to prevent them being shed or bumping against the animal's neck every time it moves. Those who study large mammals usually have rules of thumb, such as leaving room for three fingers or a hand to slide under the collar. To reduce the risk of chafing on the neck, the collar should be fastened at the side, with any metal fittings covered or at least smoothed on the inside.

Adjustable collars for small mammals are attached in several different ways. With closed loops, forceps are used to tease the double folds of wire outwards to the right circumference (Fig. 5.4), after which the heat-shrink tubing is rubbed with a portable soldering iron, or a penknife blade heated with a cigarette lighter. Slip a piece of thick paper or thin card under the

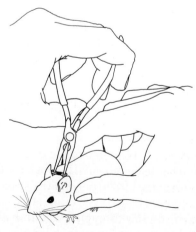

Fig. 5.4. Using fine pliers to tease apart the "sliding-8" and thus tighten a tuned loop collar on a small rodent.

5. Tag Attachment

Fig. 5.5. Putting on a squirrel collar. Don't get bitten!

tubing to insulate the animal from the heat. With an open wire loop, push some heat-shrink tubing down one wire before the insulation is stripped from the ends (Fig. 4.6). After joining the bare wires, shrink the tubing over them to strengthen the loop and to give the bare wires some protection against corrosion. Collars with fine brass strip are sometimes closed in a similar way, by folding the brass ends over each other before using heat-shrink tubing to lock them in position (Fig. 4.9A).

Large tuned-loop collars are usually closed with screws. You can avoid catching the animal's fur in the join by sliding paper or card under the collar while the screw is inserted and tightened. Squirrels and some other species can be held round the thorax while a cable-tie collar is slipped over their head. Hold the collar upside down, in order to avoid getting bitten and so that you can grip the end of the cable-tie in your teeth to tighten it (Fig. 5.5). Use wire snippers to cut the end off the tie, rotate the collar round the neck to its correct position, and release the animal. Squirrels which are gripped too tight during tagging may go limp, in which case they should be rested loosely with their feet on the ground until they recover. If this is not done at once, they stop breathing and die. About 10% of squirrels also roll up on the ground after release, trying to get the collar off with their hind legs. If left by themselves, however, they run off within a few minutes, and there is no evidence of this behaviour recurring.

F. Implanted and ingested tags

Inhaled methoxyfluorane and intramuscular ketamine (Ketalar, Vetalar) or xylacine (Rompun) have been used successfully as general anaesthetics for implanting tags in mammals and reptiles (E. Smith, 1980; Smith and Worth,

1980; Eagle et al., 1984; Madsen, 1984), tricaine methylsuphonate in water for amphibia (Stouffer et al., 1983), and MS-222 (Winter et al., 1978), benzocaine (Laird and Oswald 1975), or 2-phenoxyethanol (Johnsen, 1980) for fish. Korschgen et al. (1984) preferred local anaesthesia with 2% lidocaine hydrochloride for abdominal implantation in diving ducks, after initial tests showed that this caused no more struggling than a general anaesthetic.

Although one of the earliest small mammal tags was a subcutaneous implant for *Peromyscus* (Rawson and Hartline, 1964), recent studies of sea-otters and beavers have reported the expulsion of subcutaneous tags, leading to infection and death of the animals. No such problems occurred with intraperitoneal implants in these species (Garshelis and Siniff, 1983; Davis et al., 1984), or in the long-term tagging of several other mammals (Melquist and Hornocker, 1979; Philo et al., 1981; Eagle et al., 1984). Intraperitoneal implantation can also be recommended because tags weighing 10% of bodyweight had no affect on litter sizes in 9 g *Peromyscus* (H. Smith, 1980).

The peritoneal cavity is usually accessed by a mid-ventral incision to minimize bleeding. However, tags which are sutured in place (e.g. for ECG recording) seem less prone to expulsion if the incision is slightly to one side (Folk and Folk, 1980). Healing may also be more rapid if a lateral incision is used to keep the weight of the tag off the wound. If you may want to withdraw the tag later, a loop or socket at one end will provide something to grip and thus aid removal through a small incision. Tags are typically inserted in semi-sterile conditions, using sterile instruments and gloves, and with the animal's skin cleaned with disinfectant and alcohol but not shaved or plucked over a large area. Place tags in a surgical antiseptic solution and rinse them with sterile water before insertion. Keep animals for one or two days before release, to ensure that they do not attack the sutures or show other signs of distress. Resorbable sutures should not be used for implants in fish, because healing is relatively slow (Winter et al., 1978).

Sensor leads are often threaded for some distance under the skin. A flexible cannula can be used to make a channel towards the lead's point of insertion, so that the lead can be inserted along the cannula from its point and left in place as the cannula is withdrawn (Sawby and Gessamen, 1974). Similarly, flexible whip antennas can be inserted under the skin to provide a long radiator close to the animal's surface. This is a popular approach on fish, where the radiation loss from fish to water is only 10% (Winter et al., 1978).

Ingested tags have been used to monitor internal temperatures (Mackay, 1964; Osgood, 1970; Swingland and Frazier, 1980), and for radio-tracking species which are difficult to tag externally (Fitch and Shirer, 1971; Brown

5. Tag Attachment

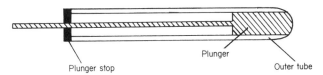

Fig. 5.6. A plunger for inserting tags into fish stomachs. The plunger forms a smooth surface with the outer tube, and is inserted first without the tag. The plunger is then withdrawn, and used to push the tag through the tube. Redrawn, with permission, from Solomon and Storeton-West (1983).

and Parker, 1976; Madsen, 1984) or where implanting is unacceptable. The tags can be moistened and inserted into the stomach with the aid of a plunger. To avoid damaging the delicate gut of salmonids, the cylindrical tags described in Chapter 4 (IV.I) are best inserted through a special tube, which presents a smooth, curved anterior surface when the convex-ended plunger is at its end (Fig. 5.6). The plunger is withdrawn when the tube's tip is in the right place, and used to push the tag down the tube and gently out into the stomach (Solomon and Storeton-West, 1983). If released in a pool with good lies, salmon usually remain there overnight, but if released in rough water between pools they often run upsteam for several kilometres and through several pools before resting. The rate of tag expulsion varies from species to species: salmon seem to retain the tags for weeks, whereas sea-trout often regurgitate them within 4–7 days (Solomon and Storeton-West, 1983). To encourage tag retention by snakes, a cord can be tied loosely round the abdomen just in front of the tag for a few days (Fitch and Shirer, 1971).

Fish with ingested tags are best marked with an external reward tag too, to improve the chance of tag recovery and reduce the time wasted on "lost-tag" searches if the fish is caught by an angler or other fisherman.

IV. TAG ADJUSTMENT AND DETACHMENT

Biologists working on mammal population dynamics are faced with the problem that much mortality and dispersal occurs before the animals reach full size, so that collars which are small enough to remain on juveniles will strangle them when they become adults. One solution is to use small ear tags (Serveen *et al.*, 1981) on the young animals. Even if the tags are eventually shed, they may first help to capture the full-grown animals for collar attachment. Another approach has been to use expandable collars. One ungulate collar design is a sliding loop, with a counterweight to keep it tight round the animal's neck, but the use of this design is restricted by the risk of entanglement in vegetation. Another expansion system has pleats in the collar material, held together by threads which break under pressure as the neck grows. Although Cochran (1980) commented that expandable collars are seldom free from problems, such as a tendency to hang low and accumulate ice masses in very cold weather (Clute and Ozoga, 1983), recent designs have proved satisfactory for bears (Strathearn et al., 1984) and bobcats (Jackson et al., 1985).

The bobcat tags also solved another problem with collars and harnesses, in that they were designed for ultimate safe detachment. Such detachment is most easily achieved from aquatic species. For instance, if the fish tags described in Chapter 4 are made with a keel, they can be attached by a biodegradable suture which passes through the hole at one end, through the back at the base of the dorsal fin and along a retaining plate, before passing again through fish to the other end of the tag (Plate 9). Since fish skin is delicate, it is important to sew through the musculature as well as the skin (Priede, 1980). Attachment is rapid using a suture on a ready-mounted

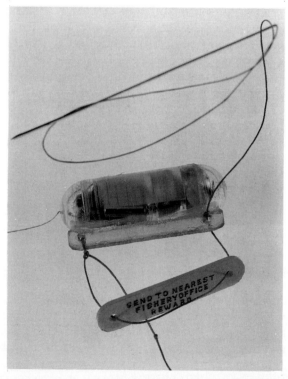

Plate 9. A fish tag for mounting just below the base of the dorsal fin, with suture for sewing through to the address label on the other side of the fish. (Photo by D. Solomon & T. Storeton-West.)

needle (e.g. Ethicon W9415, W9418), and the tag sheds after 20–30 days. Although the sutures sometimes tear out, the injury is slight and heals in due course (Solomon and Storeton-West, 1983).

The sea is an ideal environment in which to use biodegradable materials for tag detachment. Turtle harnesses have been released by using adjacent links of aluminium and stainless steel, which form a corrosion cell (Ireland, 1980), and rubber collars on sea-otters have been released after 2–3 months through the corrosion of wire staples (Loughlin,980). Radio tags for satellite tracking have been mounted as a saddle at the base of dolphin dorsal fins, using a 6.4 mm biopsy needle to bore a hole for a retaining bolt through the fin. The stainless steel bolt, covered with a sleeve of biocompatible nylon, was inserted as the needle was withdrawn to reduce bleeding, and used with a magnesium nut which would eventually corrode to release the tag (Butler and Jennings, 1980).

Biodegradable sutures can be used to release harnesses from aquatic animals, but degrade less satisfactorily on dry land. Some projects have used elastic bands, latex harnesses (Godfrey, 1970) or other perishable materials to release tags from birds, but this often leads to premature tag loss. However, harness straps of 4 mm wide clothing elastic were satisfactory for three- to five-month tags on woodcock, and had been shed by most of those recaptured a year after marking (Hirons and Owen, 1982).

In other projects, tagged animals can be retrapped or drugged relatively easily to remove or change their tags. With this in mind, Bertram (1980) recommended that those tagging large mammals should:

(i) always carry a spare tag, to replace a dying tag or to test receiving equipment
(ii) routinely record the signal pulse rate of tags, to detect the slowing which precedes cell failure
(iii) tag more than one animal in a social group, in case one tag fails, and
(iv) deal gently with all animals, because fear would make them harder to recapture and to observe.

These recommendations hold, in whole or in part, for all wildlife radio tagging.

6
Radio Tracking

I. FIRST PRINCIPLES

Radio waves behave in much the same way as light. They are subject to reflection, refraction, diffraction, interference and polarization, and their intensity diminishes with distance from the source. In line-of-sight transmissions, from a flying bird (or received in an aircraft), the signals roughly obey the inverse-square law: their strength is reduced by 75% if the range is doubled. In ground-to-ground transmissions, however, the rate at which signal intensity declines may be substantially modified by absorption in vegetation, and by phase differences between direct and indirect (e.g. reflected) waves.

Radio waves can be reflected quite strongly by cliffs, hillsides, woods, buildings, and even individual rocks and trees. This is sometimes useful, because reflected waves may be detectable when direct waves are blocked. However, reflections are more often a nuisance, giving a false impression of a tag's bearing.

Both reflection and refraction are involved in the propagation of signals from underwater tags. Rays are totally reflected back from the surface if their angle of incidence is more than 6° from the vertical, and rays which pass through the surface are refracted to an increasing extent as their incident angle diverges from the vertical (Fig. 6.1). As a result, a much stronger signal is obtained at a given distance in the air straight above a tagged fish than at the same distance away at water level: elevated antennas and searches from aircraft are especially effective when tracking fish.

Diffraction, which allows electromagnetic radiation to be detected slightly off-line from the source round the edge of an impenetrable object, can give rise to interference effects around tree trunks and thus make it difficult to get accurate bearings on woodland animals. This effect, and reflection from the trees, hinders direction finding not only within woodland, but also for signals emerging from the wood into a "Fresnel zone" which extends for about 20 wavelengths outside an abrupt woodland boundary.

Interference, occurring when signals reach a receiver by different path lengths and are therefore out of phase, produces very noticeable effects

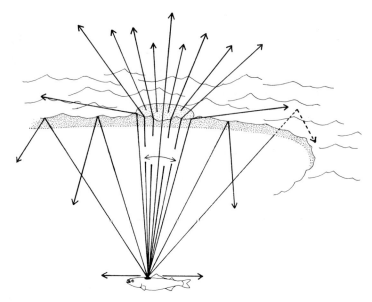

Fig. 6.1. Radio waves from a fish tag are reflected back under the water if they are more than 6° from the vertical. Some waves which pass through the surface are refracted towards the horizontal, but reception is best from overhead. Redrawn, with permission, from Priede (1982).

during radio tracking at 100–250 MHz. At 150 MHz, for instance, the wavelength is 2 m. When waves reach the receiver by pathways which differ by 2 m they reinforce each other to produce a signal peak. On the other hand, if the pathways differ by 1 m the signal tends to a null. Peaks and nulls of this type are often noticeable when approaching tags on animals in trees, if signals on a direct path from the transmitter interact with those reflected off the ground: the signal is strong in the direction of the transmitter at one point (Fig. 6.2A), but a few paces later becomes weak in that direction (Fig. 6.2B) and may then be strongest well away from the tags's true bearing. Holding the receiving antenna with its elements vertical can help to find the true bearing again, because the vertically polarized signal is least strongly reflected from the ground. In general, however, Yagi receiving antennas should be held with their elements horizontal in woodland, even though the bearing accuracy is then somewhat reduced, because the trees produce so much reflection and diffraction interference in the vertical plane.

II. MAKING A START

Before starting to track fast-moving animals, it is wise to gain some experience finding stationary tags. The first step is to place one in the open,

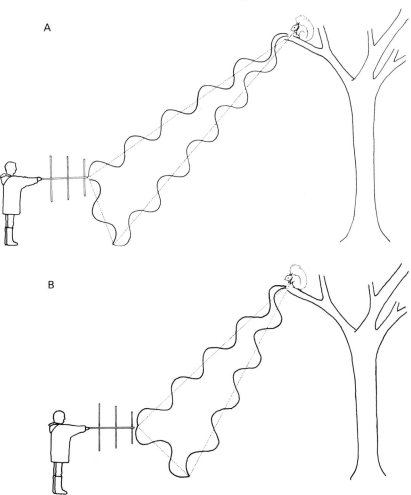

Fig. 6.2. Interference from reflected waves. At *A*, the direct and reflected waves are in phase and produce a reinforced signal in the direction of the tag. The tracker then walks forward a couple of paces, such that the direct and reflected waves are out of phase (*B*). The result is a minimal signal from the tag direction, with stronger signals from other directions.

at the height and orientation it would have when a tagged animal stands on the ground. Put the tag down near a climbable, isolated tree if your study species uses these. Take your receiving equipment 50–200 m away, connect up the receiving antenna and earplug or earphones, and switch on the receiver with the gain set low. If the receiver has a separate volume control, turn this to about half the full setting. Hold the antenna where it will give the strongest signal: this is with the boom of a Yagi or the cross-piece of an H-Adcock or the plane of a vertical loop pointing towards the transmitter, but with a horizontal dipole at right angles to the direction of the transmitter. Then turn up the gain until you can easily hear the pulsed signal from the tag. If there is a lot of background hiss, try a slightly lower setting of the volume control. Leave the volume control at this "comfortable" setting, and make any further sensitivity adjustments with the gain control.

A. Taking bearings

The polar diagram for a three-element Yagi (Fig. 2.2B) shows a marked signal peak along the line of the boom, with the shortest element (the director) towards the signal source. There is a smaller peak in the reverse direction, and relatively small peaks to the sides. Holding your Yagi with its elements vertical, turn in a circle and notice how the strong signal towards the transmitter drops off quite sharply as the antenna swings more than about 30° to the side, and then picks up again, but to a lesser extent, with your back to the tag. To take a bearing, find the peak signal position and turn down the gain until you can only just hear the tag (Fig. 6.3.A). Swing the antenna slowly to the side until the signal disappears (Fig. 6.3B), and then swing back until you can just hear it again, and note a landmark along the line of the antenna boom (Fig. 6.3C). Repeat this on the other side (Fig. 6.3D). The tag bearing is the line that bisects the angle between these two directions. Now rotate the antenna boom so that the three elements are horizontal, and repeat the process. You may notice that the null directions on each side are slightly further apart than before, or perhaps slightly more difficult to find, but you should obtain a very similar bearing for the tag. Since horizontally polarized signals often give a more diffuse peak, it is usually better in open country to take bearings with the antenna vertically polarized. If you measure the bearing with a compass, hold this instrument as far as you can from the receiver and headphones, which have weak magnetic fields of their own.

The polar diagrams of loop, Adcock and dipole antennas differ from those of the Yagi in having two opposite peaks of similar strength in their reception (or transmission) pattern (Fig. 2.2A,C,D). The transmitter bearing is usually measured through one of the intervening nulls. Since the opposing peaks and nulls are symmetrical, it is harder than with a Yagi to distinguish the true bearing from the back bearing, in the reverse direction. With practice, however, this ambiguity can often be resolved.

Hold a loop antenna vertical and quite close to the front of your body. If the loop is mounted on the top of the receiver, the receiver should be horizontal at the nearest side of the loop, like the handle of a tennis racket (Plate 10A). Turn in a circle, and note how there is a broad peak in the transmitter direction and in the reverse direction, with quite sharp nulls as

Plate 10. Taking bearings with a loop antenna. The loop is held to one side to find the signal peak, and the null bearing taken looking through the loop (see text for details).

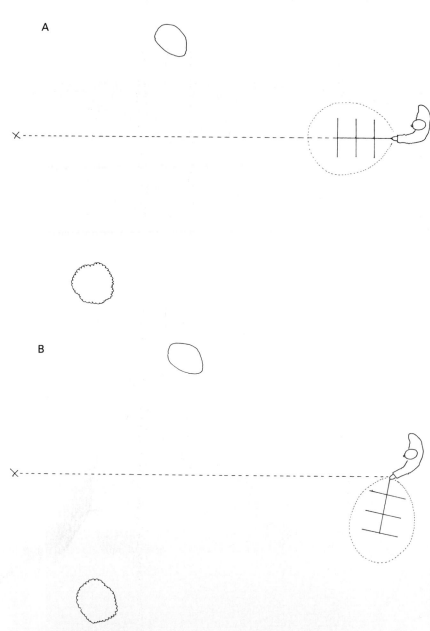

Fig. 6.3. Taking a bearing with a Yagi antenna. First find the direction with the strongest signal (*A*), making sure that it is not a reflection or back-bearing. Swing the Yagi to one side until the signal is no longer heard (*B*), then swing back until it is just detectable, and note a landmark in that direction (*C*). Repeat this procedure on the other side (*D*), and the tag direction is the bisector of the angle between the two landmarks.

C

D

you face at right angles to the tag. The signal should be slightly stronger when you face towards the tag, because waves are then reflected off your body onto the antenna, whereas you provide some shielding for the antenna when you face in the opposite direction. Now face in the direction of the strongest signal, looking through the loop as you hold it vertically (i.e. with the plane of the loop at right angles to the peak signal direction). Turn slightly to your left and to your right, but stop when you find the signal null: the tag is then on a bearing straight through the centre of the loop (Plate 10B). If the loop antenna is fixed on your receiver, you can mark a line on the receiver casing, through the loop axis, as a convenient sight-line for bearings. Unless the null is quite sharp, it may be best to turn down the gain until no signal is detectable at the null, and then find the directions in which the signal intensity increases markedly on either side: the bisector of these bearings points at the tag, as with the Yagi antenna.

If the boom of an Adcock antenna is extended as a hand-grip, you can use it in the same way as the loop antenna to estimate the direction of the transmitter, by turning in a circle with the elements vertical in front of you. Rather than hold the antenna across your front to find the null bearing, it may be easiest to hold it out to one side. If the antenna is well-tuned, there should be a very sharp null with the boom at right angles to the line from the transmitter. If you take a line along the boom (or by lining up the vertical elements), the tag is on a bearing of 90° to this line (Fig. 6.4).

The signal received by a dipole is strongest with the antenna at right angles to the direction of the transmitter. The length of a 104 MHz dipole prevents

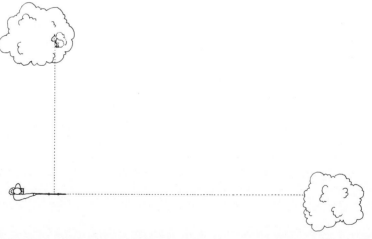

Fig. 6.4. Taking a bearing with an H-Adcock which has an extended boom. The boom is used like that of a Yagi to find the centre of the null between the two signal peaks (see Fig. 2.2C), and the tag bearing is then at 90° to the direction along the boom (on either side).

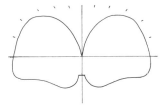

Fig. 6.5. Using a horizontal dipole, the tag bearing can sometimes be discriminated from its back-bearing by tilting the antenna. With the dipole tilted at about 15° to the horizontal (Plate 11), the sharpest null is in the direction of the tag.

the body being used as an effective shield to help determine a tag's true bearing. However, if you tilt the antenna at about 15° to the horizontal and rotate it slowly above your head (Plate 11), you may notice that the null tends to be sharpest with the upward pointing end of the antenna towards the tag (Parish 1980, Fig. 6.5). Take the null bearing along the line of the antenna.

B. Estimating tag distance and position

After discovering how to determine the direction of your stationary tag, it is worth investigating the effects of tag distance and position on the signal strength. Note the gain required to just detect the signal where you are, and walk slowly towards the tag with the receiving antenna held for peak reception. The signal will increase in strength quite rapidly as you approach, and you will probably need to reduce the gain in order to distinguish the direction of the strongest signal. You will eventually reach a point at which you can just detect the signal with the minimum gain setting. With a small-mammal collar this distance may be no more than 10 m, but it could be more than 100 m from a powerful two-stage tag.

Walk round the tag, keeping at this minimum-gain distance. Unless the tag antenna is a vertical dipole, you will notice the signal strength change, reaching a peak when you are sideways-on to a loop or a horizontal whip. The signal strength will be least when you are at the axis of a tag's loop antenna, or end-on to a tag's whip antenna. If your study species generally stays on the ground, note the maximum distance at which you can hear the signal with the receiver gain set at minimum: if you can hear a tagged animal with the minimum gain setting in the field, you will be within that distance of it.

If your study species climbs or perches in trees, put the tag 4–5 m up the nearby tree, with the same orientation as before, and again find the maximum distance at which you can hear it with minimum receiver gain.

Plate 11. Taking a bearing with a dipole antenna, which is tilted to emphasize the null towards the tag (see Fig. 6.5). Reproduced from Parish (1980), with permission.

Unless your tag was originally quite high above the ground, the distance will have increased very appreciably. If you are working with animals which must not be approached within a certain distance, to avoid disturbing them or putting yourself at risk, you may wish also to note the gain setting for just detecting the signal at that distance.

If you return to the position you first used for direction-finding, you should find that the signal from the elevated tag is much stronger than when it was near the ground. Moreover, the signal strength will increase less rapidly than before if you walk towards the tag again. This effect can be useful when gathering location data on species which are sometimes in trees and sometimes on the ground. For instance, a relatively weak signal from within a squirrel's normal range, followed by a rapid increase in signal strength as one approaches, is a cue that the animal is on the ground: this is often confirmed by a very marked further increase in signal strength as the disturbed squirrel runs up a tree.

As you walk towards the trial tag in the tree, you may notice the sort of peaks and nulls mentioned at the start of this chapter, resulting from interference between the direct wave and the ground wave. Another effect which sometimes makes direction-finding difficult at close range is signal swamping in the receiver. Many people unwittingly achieve this effect by turning the volume down, instead of the gain, to prevent the signal getting too strong in the headphones as they approach the transmitter. The receiver then gives much the same audio output, with the antenna in a peak direction and the automatic gain control cutting down the input, as with the antenna in a null position and no AGC action. The best remedy is not to change the volume control after the start of a tracking session.

C. Three-dimensional fixes

For many species, radio tracking is essentially in one plane. In other cases, signal strength at a particular distance can indicate whether the animal is in a tree rather than on the ground, or in shallow rather than deep water, but no third dimension bearings can be taken to show how high or deep it is, either because the animal is too shy to be approached closely or because of refraction at a water surface. In some circumstances, however, animals can be tracked in three dimensions, to determine their height in trees, or on cliffs or buildings.

Although loop and Adcock antennas can give elevation bearings, only a Yagi is really practical for three-dimensional tracking. If you keep the boom of a Yagi horizontal while approaching the tree which contains your stationary tag, you may notice that the signal weakens as you get very close and walk past under the tag, before increasing slightly as you continue on the other side. Turn round and approach again. When you are close to the tree, sweep the Yagi through the vertical plane, with the elements first vertical and then horizonal. The signal will tend to a maximum when the antenna boom points exactly at the tag. Interference effects between the ground wave and direct wave may tend to reduce the accuracy of this elevation bearing, particularly with the Yagi elements horizontal. When you are almost under the tag, you may even get the strongest signal with the antenna pointed at the ground, probably because of reinforcement from the reflected wave (see Fig. 6.2). If the signals become confusing, try moving round the tree and checking the tag's bearing from several positions.

III. PRACTICE TRACKING

After learning to take bearings and estimate distances to a tag in the open, try to find some tags which someone else has hidden in your study habitat. The less flat and open the habitat, and the more rapid moving your animal, the more you will benefit from the practice. Equip yourself and a companion with detailed maps of the area, and have tags hidden five or more times in the places which your study species might visit: up trees, in burrows, or in streams under a bank overhang.

This exercise can be fun, especially if you use the excuse to spend a day in the country with a friend or spouse, but do allow yourself plenty of time. It may well take you a couple of hours to find each of the first tags. To avoid making the exercise too difficult, two or three tags should be detectable from your starting point, but it is a good idea to have another two or three which require a search in order to detect the signal, especially if your study animals will quite frequently go out of range.

A. Triangulation

When you first get a bearing to a hidden tag, do not immediately rush off in that direction. A weak signal could come from a tag on or under the ground quite a short distance away, or in an elevated position more than 1 km away, especially if you are using cars to practice tracking wide-ranging animals. Moreover, the direct route to the tag may take you through thickets or streams, which could be avoided by taking a less direct road or path. You should first take cross bearings to triangulate the tag's position, and plan your approach accordingly.

Try to take bearings from sites which will ensure reasonable accuracy. Ideally, you should be in the open, well clear of rocks, buildings, power lines, wire fencing, large trees or plantation edges, which might reflect or otherwise interfere with the signals. If you cannot avoid being surrounded by rocks or buildings, try to get on top of the highest one near you. At worst, obtain an approximate direction for the signal from several different positions, and then put the largest obstacles behind you when you take the bearing. Check the bearing from one or more nearby positions, with the elements both vertical and horizontal if you are using a Yagi. Vertical elements should give the most accurate bearing in the open, horizontal ones in woodland. If the signal is coming from a wood, you will get your most accurate bearing from at least 50 m (but preferably more than 100 m) outside it, not from among the trees.

To take a second bearing, move about 200 m across the line of the first bearing. If both bearings are almost identical, the tag is a long way off, and you should arrange your route to take your next bearing at least 200 m further off the line of your first bearing when you have gone several hundred metres towards the tag (Fig. 6.6). Before you set off, make sure that you are not heading in the reverse direction to the tag! If the signal strength is not obviously weaker in the opposite direction to your bearing, try to place a large obstacle, such as a ridge, rock, building, or even a vehicle between you and the presumed position of the tag: if the signal is not appreciably weaker, and especially if it becomes stronger, suspect a back-bearing.

If the angle between the two bearings is at least 20°, and you are confident that the bearings were fairly accurate, then you have a rough idea of the tag's position. You would be wise, however, to plan another bearing from off the line of the first two as you make your approach. If you are a long way from the tag, you may be wise to triangulate it again when you are closer, preferably with at least 60° between the bearings.

While you practice with one tag at a time, you may not bother to draw the bearings on a map. If you need to draw many bearings, however, you can

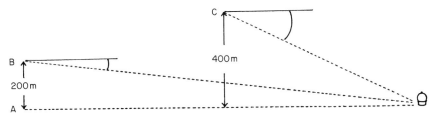

Fig. 6.6. Taking bearings to triangulate a tag. The first bearing is taken from point *A*, after which the tracker moves across the line of the bearing to point *B*. There is a small angle between the bearings, so the tag is relatively far away. The tracker then moves closer to the tag, but even further away from the line of the first bearing. This produces a relatively large angle between the first bearing and the bearing at *C*, so that the tags position can be estimated with enough confidence for a direct approach.

prolong the life of your maps by inserting them in plastic folders. Large maps can be cut to fit in loose-leaf books of such folders. Mark the bearings on the plastic with felt-tip pens, and wash them off later with a suitable solvent. When you plot bearings to obtain fixes, remember to use grid lines based on magnetic north as your reference. Many maps give these as well as the grid lines based on true north. If you lack such a map, you may need to correct your bearings for a difference of several degrees between true north and magnetic north. Another problem arises in very homogeneous habitats, such as open grassland or a large area of woodland, where you may have trouble determining your own position. Bearings are little use if you do not know their origin. One solution is to put out a grid of numbered markers before you start tracking animals (Sanderson and Sanderson, 1964).

As you move closer to the tag, and especially during the final approach, it is easiest to keep a check on the direction by swinging the antenna from side to side. This provides a sort of running average of the peak (or null) signal direction, and is especially useful in woodland or other areas where the bearing is often inaccurate. It also indicates when you start to walk past the tag, so you can adjust your course accordingly. When you are very close, swing the antenna in the vertical plane to check whether the tag is up a tree.

Once the gain has been turned down to minimum, you may have trouble deciding where a fairly loud signal is coming from, especially if you are not using a Yagi. At this point you can disconnect the antenna, turn up the gain until you can detect the signal again, and use the receiver itself to "hunt the thimble": the signal strength increases as you move the receiver closer to the tag. If you can not hear any signal without the antenna, a straightened paperclip or other short wire can be inserted in the centre of the antenna socket to increase the gain a little.

While you practice locating hidden tags, it may be wise to attach a streamer or other visible marker to them. After all, you would probably spot the tagged animal itself, or not wish to go closer, if you were close enough to detect the tag without the receiving antenna. Nevertheless, you may sometimes have to use the receiver alone to find shed tags, or dead animals which have been cached by predators. If this is a frequent occurrence, you can make a loop of about 15 cm diameter in the end of a piece of 50 Ω coaxial cable, which is threaded through a short length of plastic piping as a handle. This can be used as a loop antenna over short distances. The peak in reception as the loop is moved to surround a tag can even be used to locate unseen tags under murky water (Solomon and Storeton-West, 1983).

B. "Lost" tags

If you cannot hear one of the hidden tags, gain height. Raising the receiving antenna only 4 m above flat ground can double the reception range. The best vantage points are sharp hilltops or ridges, tall buildings and look-out towers (Plate 12), but climbing on rocks, trees or vehicle roofs often makes all the difference between success and failure to detect a faint signal. If you are going to study wide-ranging animals in flat terrain, it is well worth having a hand-held mast of aluminium tubing (e.g. a drey-poking or pigeon-lofting pole), with a clip at the top for the antenna. You might also consider attaching a telescopic mast to your vehicle (Section V.A).

Once you have lofted yourself or the antenna, or both, hold the antenna above your head and swing it slowly round the horizon. You need only swing a loop, Adcock or dipole antenna through 180°, but a Yagi should be swung through a complete circle twice, once with its elements vertical and once with them horizontal. If you are using a Yagi with more than three elements, or a tag with a leisurely pulse rate, it is especially important to swing the antenna slowly. Otherwise you may swing right past the tag bearing during the period of silence between signal pulses.

If you still cannot hear a hidden tag, then check the battery strength and antenna connections of your receiver: there should be a noticeable change in the strength of the background "hiss" as the antenna is plugged in. Failure of

Plate 12. An ideal high point for detecting signals from distant tags.

antenna connection, in worn coaxial cables or at the plug centre-pin within LA-12 receivers, is probably the single most common cause of poor reception. If your equipment is working properly, then it seems that your helper is being really cruel: you can either start a mobile search (Section V) or give up and go home!

IV. SIGNALS FROM TAGGED ANIMALS

A. Movement cues

When you listen to a tag on a moving animal, the signal strength often fluctuates quite noticeably. This amplitude variation results partly from changes in the orientation of the tag's antenna, which can direct peaks or nulls at the receiver, and partly from changes in the signal path, for example when the animal moves behind obstacles. The extent of these variations depends on the tag design and the habitat. Thus, there may be little variation in the signals from tags with vertical whip antennas in open country. On the other hand, signal volume can change dramatically from one moment to the next as an animal with a loop antenna forages on the woodland floor. Not only is there a null when the animal faces directly towards or away from you, but the signal also attenuates when the animal is behind large trees or in ditches. There may also be slight changes in signal frequency as the tag's antenna moves in relation to the animal or brushes against vegetation, especially if this is wet. These frequency changes are especially pronounced for UHF tags (Lawson et al., 1976).

The flight of birds or bats can produce similar, if less marked, changes in signal amplitude. Accipiters, for example, frequently change hunting perches in open country by dropping and flying inconspicuously close to the ground, before swinging up into the crown of another tree some distance away. The steady signal from a perched hawk drops to a weak, fluctuating signal as it flies near the ground, finally increasing to a steady signal again at the end of the flight. On the other hand, hawks hunting in coniferous woodland often rise to fly above the canopy between hunting perches, which produces a characteristic increase in signal volume during flight. The latter increase is most marked if the receiving antenna is horizontally polarized, and thus receives a stronger signal from the horizontal antenna on a flying hawk than from the more vertically oriented antenna on a perched hawk (Widén, 1982). Tail-mounted posture-sensors give an even more precise indication of flight, by changing to a rapid pulse rate as the tail becomes horizontal, provided that one discriminates between this rapid, fluctuating signal and the rapid, steady signal from an incubating hawk (Kenward,

1985). This is by no means the end of the story, however, because soaring produces a characteristic signal too, which waxes and wanes regularly in intensity as the transmitting antenna alternately directs peaks and nulls at the receiver. With the addition of bathing, which is indicated by marked frequency fluctuations as the antenna grounds in the water, a total of six behaviours can be interpreted from tags tail-mounted on live hawks: perching, soaring, other flight, bathing and, with posture sensors, feeding and incubation.

The converse of the activity cues are the indications that a tagged animal is dead. The signal then remains steady; it is often relatively weak because the corpse is lying on the ground, and it does not move between tracking periods. Tagged goshawks have a tendency to clutch the ground with their feet as they die, so that they give the same rapid signal as an incubating hawk (albeit much weaker), whereas other bird species often die on their backs.

You may well only be able to interpret the different signals if you can first watch some tagged animals. The opportunity to do this, together with the need to check that a tag has no marked adverse effect on its wearer, are two good reasons for tagging one or two captive animals before starting fieldwork.

B. Tracking moving and distant tags

Sometimes tagged animals can be detected throughout their range from one or more vantage points. In other cases, however, they have to be followed to different parts of their range if they are to be detected. This can mean that tracking is quite difficult during the first few days of monitoring each individual. Thereafter, however, you will not only have discovered the points from which you detect signals most easily, but also have gained an understanding of the individual's habits, so that you can often predict where it will go next (Macdonald, 1978).

Before you reach this stage, you will probably face the problem of signals which go out of range while you are tracking them, especially if you are studying birds or bats which forage several kilometres from their roosts. A typical example would be for such an animal to disappear for an hour or two, and then be back at its original position when you return from a long and unsuccessful search. The best policy when following such animals is not to let them get too far away. It is not a good idea to sit on a nearby hilltop and wait until the signal disappears before following it, because the animal may then be far away in a well-shielded gully. Keep close to the moving tag, so that you can most easily estimate its direction as it moves away from you. If the signal fades sharply and disappears at any point, it probably means that the animal has gone over the brow of a hill: move quickly to check the next valley. If the road pattern obliges you to make detours, stop to take a quick bearing at one or two convenient high points on the way, but aim to get back close to the animal as soon as possible.

You may lose the signal in the end, especially if you are relatively inexperienced. If the animal is on a foraging trip from which it will soon return to a known site, then your best option is probably to return there and wait for its next excursion, using the intervening time to study the map and decide where it might have gone. If the animal is unlikely to return to a central place, then you should start searching for the signal from nearby high points. Aim to work along the last known flight line of the bird or bat, although you may try first from an unusually good vantage point off to the side or even behind you. If possible, however, stay within the minimum likely detection distance from the flight line. If you are in flat country, check again for a signal at about twice this distance from your starting point.

At some point in your search you may pick up a very faint signal. This might last for only a few moments, while transmission conditions are optimal for a very distant tag, so get the peak signal direction as quickly as you can. Check two or three times that you are not using a back-bearing: if the signal strength is fluctuating, keep swinging round to check until you are quite sure of the direction. Move in the tag's direction to the next good high point, but bear in mind the need to cover for weak transmissions from the intervening area, perhaps with one or two quick stops on the way. If you have no further contact after going further than the maximum likely detection distance, and the tag appeared to be still when you last heard it, then you would be wise to return to your starting point and check again that you are not following a back-bearing.

C. Disappearances

If an animal with which you are familiar cannot be detected from any of your usual vantage points within its range, and you have checked that you are receiving properly (e.g. by listening to another tag), there are three possible explanations. Firstly, the tag may have stopped transmitting. If it was not near the end of its cell life, and you have detected no slowing of the signal pulse rate which could precede cell failure, then this is relatively unlikely. You should in any case check the second possibility, that the animal is still within its range but transmitting a very weak signal, perhaps because of antenna failure, or because it has been buried by a predator. So before looking further afield, it is worth looking systematically for weak signals throughout the animal's known range. If a broken whip antenna is suspec-

6. Radio Tracking 133

Plate 13. Radio tags help to find dead animals quickly, and thus to determine the causes of death. This squirrel had been cached underground by a stoat.

ted, visit the the animal's lairs, nests or roosts at a time when it is likely to be there. It may also be worth visiting predator lairs or other possible cache sites (Plate 13), and areas where you would expect poor reception from a tag on the ground: if a tagged animal lies dead in a ravine, dike or stream bed, there may be very poor reception unless you stand on the edge.

If you still have no signal, you can either write the animal off, or undertake a more extensive search away from its normal range. Whether you search on foot or by vehicle, there are several ways to increase your chance of finding the lost tag. Aim to search at a time of day when the animal's behaviour will ensure a good reception range. For instance, if your species feeds by day on the ground but spends the night in trees, search for it after dark. On the other hand, if the animals rest in burrows, search for them during hours of activity. Conduct your search as soon as possible after you lost the tag, not only because the animal may get further away the longer you leave it, but also because you will probably receive the signal best while it remains alive. Decide in advance on the minimum distance at which you are likely to detect the tag, and plan your search accordingly, using a map on which contour lines and other features indicate the best listening points. You may wish to start your search along leading features, such as hedges, valleys or lake shores which the animal may have followed, especially if its last position or previous experience with other individuals suggests that these are likely routes, but do not totally neglect to check the other directions. If you have a large area to cover, give due consideration to searching by aircraft. Large areas can usually be covered more thoroughly and much more quickly by air, especially if the terrain is rough and inaccessible. Even if you are not costing your own time, the cost of hiring a plane may not be very much greater than the high mileage cost of conducting a tight search pattern on the ground.

V. MOTORIZED TRACKING

A. Road vehicles

If you are routinely tracking over a large area in a vehicle, you will soon get to know those vantage points which give good reception and furnish accurate bearings. If you are tracking relatively few animals, and weather conditions are usually good, you may simply climb on the vehicle roof or up to a high point near the road to enhance reception. However, if you are working with many tagged animals, or at night and in bad weather conditions, it is better to use a vehicle with an antenna which can be raised and rotated from inside. You can then sit in relative comfort to take your bearings, with easy access to maps and notebooks, and without the risk of falling from an icy car roof in the dark.

Vehicles to be used as antenna-mounts for mobile tracking should have easy access from the driving seat to the rear of the cab, preferably with an aisle which is unobstructed by a transmission cover, and plenty of headroom inside. This not only ensures easy access to the antenna, but also allows it to be raised periscope fashion through at least 1 m. Minibuses make good mounts for antennas, but Land Rovers and other vehicles can also be used.

Steel tubing is suitable for a simple antenna mast. This can be slid up and down in a second or two through a well-lubricated bush, which is mounted on ball-bearings in a short metal tube welded through the centre of the roof. The base of the mast contains a hand-grip for raising, lowering and rotating it. There should also be a device for locking the raised base to the rotatable bush, such as a pin which is slid through matching holes in the mast and bush (Fig. 6.7). When lowered, rotation can be prevented by cutting a vertical slot in the mast's base, which engages a pin welded across a short tube on the floor.

In the United Kingdom, the Ministry of Agriculture, Fisheries and Food and the Game Conservancy have used vehicles equipped with lightweight 5–11 m telescopic masts (Solomon and Storeton-West, 1983; J. Reynolds and S. Tapper, personal communication). Using a compressed-air cylinder, a 5 m antenna elevation can be achieved in about 10 s. Such a system costs about £600 from Clark Masts (see Appendix I for address). The gain in reception from telescopic masts must be balanced against their cost and time requirements, compared with masts which can be raised through a shorter distance by hand. The balance most favours the telescopic type if the terrain is flat with a poor road network.

The most popular mast-top antenna is a six-element Yagi, with its lead passing down the centre of the mast. Vertical antenna elements give the

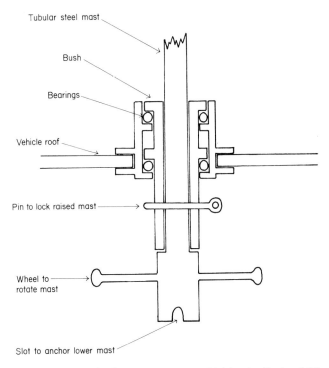

Fig. 6.7. Through-roof mounting for an antenna mast which is raised by hand. The mast must be firmly anchored in both the raised and lowered positions.

greatest accuracy in relatively open country, and also ensure the best reception range to tags with vertical whip antennas. For even greater range and bearing accuracy you can use a null-peak system with twin vertical Yagis, but this increases the loading on the mast. Remember, too, that assumptions of high bearing accuracy are only justified by a relatively "uncluttered" signal path and reception position. Moreover, vertical Yagis extend quite high above the vehicle in the lowered position, which makes them impractical in some habitats because of overhanging trees. The Yagi may then have to be used with its elements horizontal, and no more than some 50 cm above the roof when lowered. The lowered antenna should have padded supports if it will be used on rough tracks. An even more sophisticated approach is to use a Yagi which is stored horizontally, but has its boom rotated to bring the elements vertical when it is raised.

If you are working on a very limited budget, bearings can be taken with a compass rose fixed to the base of the mast. In this case, you should mount a compass inside the vehicle, at a point which is little influenced by the vehicle's own field (e.g. on the windscreen), and reset the compass rose to

Plate 14. A map table in a mobile tracking vehicle, with the display from a repeater compass (see Plate 2).

the vehicle's heading each time you stop to take a bearing. It is much easier and more accurate, however, to mount a repeater compass at the top of the mast, with its display on a mast-side map-table in the cab (Plate 14).

Before you start tracking, you should of course zero your compass, and check that neither the compass electronics nor the vehicle lighting interferes with the receiver. For instance, strip lights contain a cold plasma which can produce very appreciable interference for a sensitive receiver nearby. Plan your route in advance so that you stop at sufficient points to cover the terrain, but travel as short a total distance as possible. During tracking, remember to rotate the antenna alternately clockwise and anticlockwise, to prevent the cable twisting up, and to lower the mast before moving on. Avoid raising masts under power lines or other wires. This mistake, which is easily made at night, not only gives erroneous bearings but can be fatal!

B. Aircraft

The ideal aircraft for radio tracking are high-winged monoplanes, such as the Cessna 172, because the wing-struts provide excellent antenna mounts. Although loop and whip antennas can be mounted under the fuselage of other aircraft, attachment to wing-struts is usually not only easier, but also keeps the antennas furthest from the detuning influence of the wing or fuselage surfaces. In flying a search pattern over a large area, the reception range on each side of the aircraft determines how far apart each parallel search path need be, and thus how much flying is necessary to cover the area. The greatest sideways range is obtained by mounting a vertically polarized Yagi antenna under each wing, pointing straight to the side and 15–30° downwards from the horizontal (Gilmer *et al.*, 1981). To reduce drag and avoid any risk of over-stressing the struts, most aerial tracking is done with three- or four-element Yagis, which should ideally have thin elements and a streamlined boom: the drag on a pair of robust antennas which are normally hand-held is enough to slow the aircraft by up to about 5%.

For optimal reception, the upper Yagi elements should be at least 15 cm below the wing surface and 30 cm forward of the leading edge (Gilmer *et al.*, 1981). Antennas can be mounted like this on a boom projecting forward from the strut, but it is difficult to fix them firmly and prevent vibration. A simpler attachment, which does not greatly reduce range if the antenna is mounted a bit further below the wing, involves clamping the antenna boom close to the strut (Plate 15). A layer of dense foam rubber is glued to the clamps to avoid damaging the strut, and the antenna leads are taped to the strut or wing surface to prevent possible damage due to vibration. To minimize destructive interference between the antennas, it is advisable to keep their collectors close to a whole wavelength apart, with the leads the same length from each. Feed the antenna leads into the cabin through an air-vent, or round the edge of the doors if these are rubber-clad and will not damage the coaxial cable. The leads should meet at a switch box which can connect either the left, or the right, or both antennas to the receiver. To achieve maximum reinforcement of a side-on signal, especially if the antennas are not a whole wavelength apart, you can build a balun match into this box.

Well in advance of any flying, you should check the fit of your equipment on the aircraft yourself, and ask the pilot to inspect it too. Check that you can hear a tag you have placed several hundred metres away, with and without the engine running. If your earphones do not fit well enough to reduce the engine noise, you may have to turn up the receiver volume so much that you risk damaging your ears. Check that electrical noise from the engine is adequately suppressed: this often varies from one aircraft to another, and it is worth finding which has least if you have a choice (Mech, 1983). If you have not tracked from the air before, it may be wise to put out tags in sites with normal and poor transmission characteristics (e.g. as if an animal were dead on the ground), and make a short preliminary flight to find the ranges at which you can detect them.

If you are searching a wide area for several transmitters, plan your search pattern in advance. Discuss it with the pilot to ensure that you have chosen easily recognized turning points, and that you can obtain advance permission to traverse restricted (e.g. military) airspace. If the tags cannot be

Plate 15. Side-looking Yagi antennas attached to the struts on a Cessna aircraft (left), with a close-up of the mounting bracket (right).

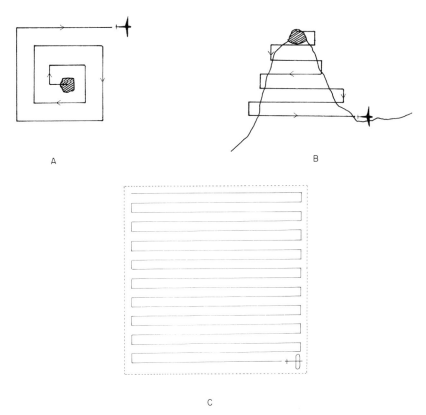

Fig. 6.8. Search patterns in aircraft. If animals may have dispersed in all directions, the search may spiral outwards from a study area (*A*). If animals must have moved in one direction, perhaps along a peninsula, search along their likely path (*B*). If the tagged animals could be anywhere within a large area, then fly a grid pattern. In all cases, the distance between adjacent search paths is twice the minimum sideways detection distance for the tags.

detected much more than about 10 km away, it is usually easiest to fly a rectangular search pattern, with the parallel transects separated by just less than twice the sideways distance at which you expect to detect a transmitter. If the animals have spread from one small area, as in the early stages of dispersal studies, you may wish to start near their origin and work outwards, either in all directions (Fig. 6.8*A*) or in the direction which the topography leads them (Fig. 6.8*B*). If they may be anywhere within an area, then you will probably need a regular grid search (Fig. 6.8*C*), although you may decide to try certain favourite sites first. For instance, goshawk dispersal is monitored on the Baltic island of Gotland by flying a grid search from north to south, but the first step in the search is to circumnavigate the coast from the seaward side, not only because dispersing hawks frequently settle in coastal areas, but also because cliffs block signal transmission inland from hawks on the ground near the shore. You may have to modify the flight path in other ways to take account of local topography. For example, signals from bats roosting in caves may only be detected as you traverse the cave entrance. You may also need to fly higher above rugged terrain than the 300–1000 m above ground level (AGL) which is adequate for tracking weak tags in lowland habitats.

For this sort of search it is preferable to use using a programmable receiver, to reduce wear on the tuning knobs (and your fingers!). Make sure that the receiver battery is fully charged before the flight, and enter the frequencies into the memory. The dwell time on each frequency should be set so that all frequencies are covered at least once while the aircraft's forward track is about two-thirds of the sideways detection limit. Thus if the sideways detection limit is set at 4.5 km (i.e. search transects 9 km apart), a scan of all frequencies should occur every 3 km. If the maximum ground speed, which is the airspeed plus the windspeed if flying downwind, is 180 km h^{-1}, then all frequencies should be scanned once per minute. This allows a dwell time of 5 s on 12 frequencies, or 10 s on 6 frequencies. If detection is better with only one antenna than with both connected to the receiver, you might decide on a cycle of 30 s and switch from one antenna to the other at the end of each cycle. However, you should not reduce the dwell time below about 3 s per frequency (and preferably not below 5 s), even if tag pulse rates are quite fast.

When you think you can detect a signal, stop the scan and switch quickly to the antenna on one side and on the other: the tag is on the side with the stronger signal. If you cannot detect the signal again, ask the pilot to circle while you listen to the antenna on the outer wing. If you still cannot detect a signal, continue flying your search grid, but mark the frequency on the map so that you can give it special attention during the adjacent transects on either side. If you have a good signal with one antenna connected, ask the pilot to circle away from that side (i.e. to the left with a signal on the right). Note the headings when the signal fades and then reappears with a similar intensity: the tag bearing is very roughly at 90° to the bisector of these angles. Another way to estimate the tag's bearing is to record your position, during subsequent scans of the frequency along the same transect, when the signal is strongest on the receiver's meter: the tag is then at about 90° to the aircraft's heading. You can get a third bearing by circling again further along the same transect (Fig. 6.9). If you only hear the signal during one or two consecutive scans, you can get a rough idea of its position from its relative strength along the next transect: if the tag was weak from one transect, and undetectable from the next, it is probably close to the first.

6. Radio Tracking 141

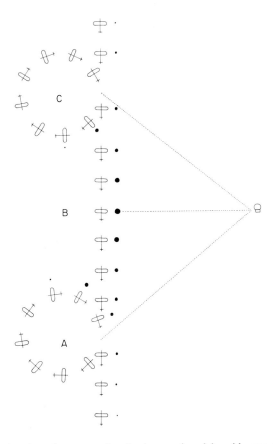

Fig. 6.9. Taking bearings along a search path when an aircraft has sideways-pointing Yagi or end-fire antennas. A bearing is taken by circling when the tag is first detected (*A*), its bearing is at right angles to the path when the signal is strongest (*B*), and a third bearing is taken by circling (*C*) before the signal is finally lost.

You may well require no more than a rough idea of a tag's position from the air, so that you can save flying costs by a subsequent check on the ground. If you want to locate the tag more precisely, however, perhaps because you are dependent on aerial searches over inaccessible terrain, then you can home in on it (Mech, 1983). Once you have obtained a bearing by circling, then head towards the tag. If the signal was relatively weak, you may lose it for a while, but it should reappear on one side or the other (Fig. 6.10*A*). Turn gently towards that side until the signal becomes equally strong from each antenna (Fig. 6.10*B*). You should then be heading directly towards the tag, and you may descend to 300 m AGL or even lower if the terrain permits. Keep heading towards the tag, reducing the receiver gain as you approach and using each antenna in turn to maintain the correct heading. As you pass over the tag the signal will peak and then start to fade again, especially noticeably if you are flying low. When this occurs, start a turn. If the signal is strongest on the inside wing as you turn, you have indeed gone past the tag (Fig. 6.10C). If the signal is strongest from the outside antenna (Fig. 6.10D), then you have gone through an interference peak and null instead of passing the tag: complete the turn through 360° and continue towards the tag. When you have definitely passed it, continue your turn to fly back towards the tag, listening carefully to each antenna in turn. It is unlikely that you will fly straight overhead, so the signal should be strongest on one side. Circle towards that side: if the signal remains on the inside antenna, you are circling the tag. If at any point on your turn the signal becomes stronger on the outer antenna, start circling to that side, until you are finally circling the tag at relatively low altitude.

If the transmitters are usually detectable in line-of-sight at least 15–20 km away, and you are looking for one animal at a time or for a small number within a relatively small radius, you may use a non-programmable receiver and a simplified search technique. This involves starting near a central point, such as an individual's last known position, and then spiralling upwards until a signal is detected or you reach, say, 5000 m (Mech, 1983). You can switch

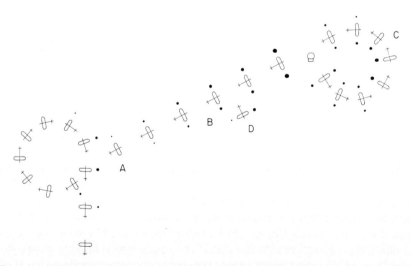

Fig. 6.10. Approaching a tag in an aircraft with sideways-pointing antennas. After circling to obtain a bearing, the aicraft heads towards the tag (*A*), keeping the signal at equal strength in each antenna (*B*). When the signal fades, the plane turns. Provided the signal remains on the inside wing (*C*), the tag has been passed, and the plane continues towards it from a new direction. If the signal is on the outside wing during the turn (*D*), the turn was made too soon and the original heading must be resumed.

from one frequency to another, or from the antenna on one side to the other, each time the aircraft's compass passes north. The higher you go, the more likely you are to achieve an uncluttered signal path from the tag. The same type of search can be fruitful for locating fish in large lakes, because the strongest signals are radiated almost vertically.

If you are tracking fish, or using an aircraft without wing-struts, you may well not be able to use a Yagi antenna system. Tags have been located fairly accurately with a vertical loop on each wing-strut, one looking forward and one to the side (Winter *et al.*, 1978). A single side-looking loop on a strut or the fuselage would probably give reasonable bearing accuracy for locating fish along a river, by flying slightly to one side and noting the position of the null between two similar peaks. Half-wave dipoles, which are best mounted horizontally and as far below any metal surface as possible, give one-third to one-half the range of a Yagi system, and are really only suitable for locating the general area of tags which are to be fixed precisely from the ground. Better gain and directionality can be obtained with an "end-fire" array of four quarter-wave dipoles on each wing-tip (Whitehouse and Steven, 1977; Whitehouse, 1980). Each antenna is one-quarter wave long and the same distance from the next, with a three-quarter wave balun match between them. The result is that signals reinforce along the line of the antennas, producing a reception peak at 90° to the aircraft axis on each side.

Anyone intending to radio-track from the air should try to obtain U.S. Fish and Wildlife Resource Publication 140 (Gilmer *et al.*, 1981), which gives comprehensive advice on the subject.

C. Off-shore tracking

Boats for radio tracking on large lakes can be equipped with a variety of antenna systems, including loops, Adcock and Yagi antennas, on fixed or telescopic masts. Aircraft-style search patterns can be used to locate tagged fish or other animals, followed by homing to the position with the strongest, omnidirectional signal. The problem is then not so much one of finding the tag, as of determining the position of the receiving vessel. This is usually done by taking bearings, with a compass or sextant, to landmarks on the shore (Winter *et al.*, 1978).

VI. DATA RECORDING

A. Radio surveillance

In radio surveillance projects, the tag is used mainly to find an animal so that it can be watched. This technique has proved particularly useful for finding nocturnal animals, such as foxes or badgers, so that their behaviour can be observed with infra-red equipment or image intensifiers (Macdonald, 1978; Kruuk et al., 1979). In other cases the tagging has enabled systematic collection of data on behaviours which are otherwise seen infrequently or with a strong observation bias. An example is work on the social behaviour and hunting of large predators and their prey (Seidensticker et al., 1970; Mech, 1980; Bertram, 1980). The behaviour can be sampled when the observer chooses, rather than on the few occasions when luck, or light cover or a particularly long chase, make it especially conspicuous. Following goshawks has not only enabled their hunting behaviour to be quantified without a strong bias towards observations in open country (Kenward, 1982b; see also Marquiss and Newton, 1982), but has also shown just how much bias could occur when predation was assessed by searching an area for kills instead of radio-tracking the hawks (Ziesemer, 1981). Searching for kills recorded all the pale-feathered pluckings where radio-tagged hawks had killed pigeons, but only one-third of the dark-feathered pluckings of pheasants, and one-eighth of the rabbit kills: the soft fur pluckings were rapidly washed down by rain, and the rabbit remains were also frequently stolen by foxes.

If animals are not to be disturbed during radio surveillance, the observer must be good not only at moving silently and inconspicuously through the countryside, but also at using the radio signal to estimate the tagged animal's distance and position. This should be triangulated when you are close enough to be accurate, but not so close that you lack adequate cover when you move to the side to take a cross-bearing. And avoid disturbing other animals, which may either alert your subject or fall prey to it as they flee from you! Unless your subject is tame and unlikely to be influenced by your approach, do not simply approach the animal until you see it, because it will probably see your movement first. Moreover, if you get upwind of a mammal it may well respond to your scent before you have even seen it. To test whether your field technique disturbs the subjects, you may be able to collect before-and-after data on their movement rates or habitat choice.

Sometimes, of course, you may have to disturb the animal in order to collect data. You may, for instance, want to make measurements at a kill. Even this may require skilful stalking, either so that you do not frighten the animal too severely by suddenly appearing right in front of it (a potentially dangerous mistake with large carnivores), or because the kill may be small and the remains difficult to find if the predator departs unseen before you get close.

In projects where tagging is being used to find and observe social animals, such as group-living mammals or flocking birds, you may well need a visual marker to distinguish the tagged individual. Wing-tags and "fin"-tags have been used for this purpose on birds, and reflective side-flashes or Beta-lights attached to radio tags on nocturnal mammals.

B. Position sampling

If you are following one animal continously, perhaps to sample its behaviour for an hour before moving to the next individual, you will probably use a dictaphone or keyboard data recorder, to avoid missing observations while looking down at a notebook. If you want to record the animal's movements, you may note its position every five minutes, say, or record the new position and time whenever it moves to a new perch or feeding site.

You should be aware, however, of a potential problem when you come to analyse these position data for habitat preferences or proximity to other individuals. Unless the animal is likely to cross many habitat boundaries or encounter many individuals with each move, the positions will hardly be independent of each other in a statistical sense: if the animal is in a 4 ha habitat block, and moves an average 10 m between fixes, it is much more likely to remain in that habitat between moves than to change.

There are several ways round this problem, but each tends to create data redundancy. You can use only one position, such as the first, from each sampling session. This wastes most of the data, however, so you may prefer to sample from within your set of fixes, at intervals in which the animal had time to cross several boundaries. Swihart and Slade (1985) have developed a test for the minimum interval between fixes which gives spatial independence in range recording. This interval is in effect the time for an animal to cross its range, and therefore provides a very conservative interval for habitat analysis. A rigorous test for data independence in habitat sampling has yet to be developed. However, you can probably use preliminary data to estimate the time taken, with 95% probability, to cross a habitat boundary. Provided that your fixes are spatio-temporally independent, you can use chi-squared tests to compare the frequency of points in each habitat with the habitat frequency in your study area, or in the animal's range. If the fixes lack independence, you can use the proportion of time spent in each habitat during each sample period, but this gives only one measure for each habitat during each period: you will require quite a large number of sample periods to test whether each habitat is preferred or avoided.

This data redundancy effect means that, if your main aim is to collect data on range use, social interactions and some other behaviours, you should consider carefully whether to take one fix for each individual as quickly as possible, and repeat this at intervals which confer reasonable independence between points, or to record all the positions for one individual continuously

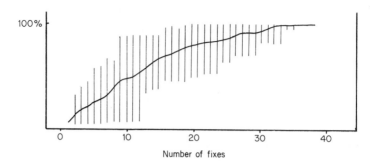

Fig. 6.11. The mean, maximum and minimum polygon range sizes were calculated for 14 squirrels as sample size increased from 3 to 40 fixes, taking 3 fixes per day. Ranges were all at or near their maximum size after 30 fixes, so standard ranges were subsequently measured for only 30 fixes.

over a long period. If it takes a long time to move between your study animals, then you will probably be constrained to period-sampling, whereas if you have dense study populations you may be wiser to opt for single-fix sampling.

Taking single fixes on a series of individuals in rapid succession enables their behaviour to be compared at similar times of day, and is also an economic way of building up range maps. The measured range size increases very rapidly when you start recording, since the first few position samples are unlikely to reveal the boundaries of the range that the animal is currently using, but as you record more positions you reach a point at which new observations are adding little to the measured range size (Hayne, 1949; Stickel, 1954; Odum and Kuenzler, 1955). Beyond this point of "sampling saturation" the animal's real range size may well continue to increase slightly, as it makes occasional excursions, but you have obtained a reasonable short-term range estimate (Fig. 6.11). Recording two or three positions a day, sampling saturation was reached after about thirty points for four species with quite different ranging behaviour: goshawk, kestrel, badger and grey squirrel (Kenward, 1976; A. Village, personal communication, Parish and Kruuk, 1982; Kenward, 1982a). The thirty-point range may therefore provide quite a useful standard for interspecific, as well as intraspecific comparisons. Range analysis techniques are considered more thoroughly in Chapter 8.

C. Accuracy

When two bearings are used to triangulate a tagged animal's position, errors can arise in several ways (Heezen and Tester, 1967; Springer, 1979;

6. Radio Tracking

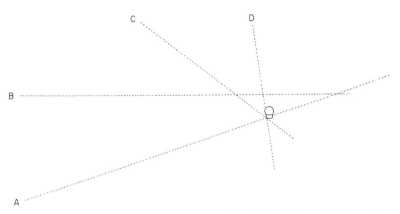

Fig. 6.12. Taking repeated bearings for accurate triangulation. The angle between the first two bearings (A and B) was small, so a third bearing (C) was taken from closer to the tag. Since these three bearings did not cross at one point, a fourth bearing was taken (D) from even closer to the tag.

Macdonald and Amlaner, 1980). For a start, a bearing may be grossly inaccurate, because topographical features have deflected the signal or caused interference in other ways. Inaccuracy may be suspected at the time, perhaps because a Yagi is not receiving a clear signal peak with lateral nulls. Moving a few metres forward or to one side often improves the reception in such cases. Even if such inaccuracies are not suspected, it is often wise to take three bearings, especially for long-range triangulation. Plot them on the map immediately, and use the centre of the triangle where the three lines meet to estimate the animal's position (Mech, 1983). If the triangle is too large, continue taking bearings (Fig. 6.12) until several coincide. Note that if you merely record pairs of bearings in a notebook, without plotting them at the time, you will have no way of checking peculiar fixes, and may well have to discard some of the data as unreliable.

Apart from topographical errors, if bearings are not taken simultaneously from two (or more) sites, there may be errors due to an animal's movement. These will be especially severe if the angle between the bearings is small (Fig. 6.13) and if the animal moves frequently. This is a particular problem when tracking birds, which can fly rapidly across country, making it vital to take consecutive bearings within 2–3 min and to take a third bearing if the bird was heard to move during this process.

As well as topographical and movement errors, which are to some extent avoidable, there are errors caused by the receiving system. Some system errors arise from misuse of equipment, through damage to antennas, or through taking readings when antennas or compasses are too close to cars or other large metal objects. These errors too are avoidable, but even the best

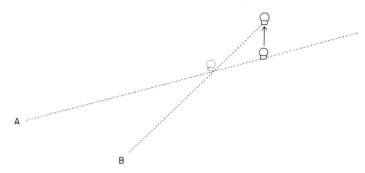

Fig. 6.13. Movement error! The animal moved some distance between the taking of bearings A and B, and its triangulated position was therefore inaccurate.

antenna has limited accuracy, which largely determines the accuracy of a skilful tracker. Antenna accuracy thus establishes the resolution of the fix coordinates. For example, if the antenna gives a bearing accuracy of about 5°, the lateral error of the animal's position is about 9% (tan 5° = 0.0875) of the distance to it. It would therefore by unrealistic to give an animal's position at 1 km with a precision greater than 100 m. If your habitat analysis requires a resolution of 10 m, then you should not take many bearings from more than 100 m. Moreover, you should remember that the size of the error polygon increases as the angle between the bearings is reduced (Fig. 6.14). Strictly speaking, it is the longest diagonal of the error polygon (Salz and Alkon, 1985) which should not exceed the diagonal of your chosen minimum grid unit. The error polygon's diagonal should therefore not exceed 14 m for a 10 m resolution.

These sources of error reduce the accuracy of your fixes, but should not introduce a systematic bias into your results. Provided that you randomly allocate (or discard) the fixes that lie on habitat boundaries, and do not assume the animal was in the most likely habitat, the errors are simply "noise": they reduce your chance of obtaining statistically significant differences, but do not bias the result. If you do not always find the tagged animal, however, then your range record may be seriously biased: the animal could well be outside the range which you have recorded so far, and possibly doing something unusual. This sort of failure to find an animal therefore tends to bias against recording large ranges, at the least, and should be avoided if at all possible.

D. Mortality recording

Mortality studies need very careful planning, with pilot work to check that tag attachment is harmless, that the tags are reliable, and that you can make

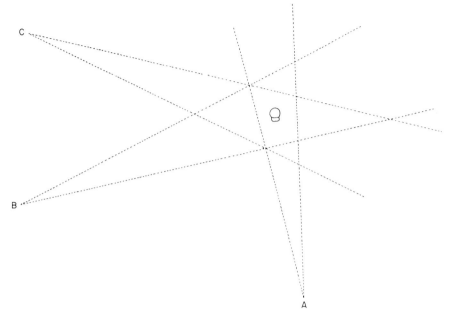

Fig. 6.14. The size of error polygons increases as the angle between bearings is reduced. The error polygon between bearings *A* and *B* is the smallest.

allowances for dispersal. The data will be easiest to process if survival checks are virtually simultaneous for all the tagged animals. If they are checked once a week, check them all on the same day if you can, or on two consecutive days. Checks should be as frequent as possible, so that mortality rates between periods can be compared by using many short-interval samples from within each period (Heisey and Fuller, 1985; see Chapter 8).

It is well worth making checks especially frequently just after tagging. This will provide plenty of data to test for any short-term effect of tagging, or for a tendency to trap animals which were about to die. Moreover, radios which fail prematurely often do so in the first few days, for example because of the voltage drop in a lithium/copper-oxide cell (Chapter 2).

It is very important to detect radio failures. You should therefore record the signal pulse rate when checking each animal, and note irregularities or changes in rate which may precede failure. Radio failures can also be confirmed by seeing the animals, or by the completion of breeding in birds whose signal has been lost. Animals "lost" prematurely will then fall into two groups: those whose tags were known to fail or which shed their tags, and those whose signals were lost unaccountably. If the resighting or retrapping rate for both groups is the same, then it is reasonable to assume that the unaccountable losses were mainly due to undetected tag failure. If there was a relatively low recapture rate for animals which were unaccountably lost, then there may have been undetected emigration, or some deaths were associated with radio failure (through impact, or destruction by a predator). If you have enough follow-up data on animals which were accountably and unaccountably lost, you will be able to correct your recorded mortality rates.

7
Fixed Stations

Fixed stations have been used mainly for radio location, for presence/absence recording, and for radio telemetry.

I. RADIO LOCATION

A. Automatic systems

The very earliest fixed stations, for example at the Cedar Creek Natural History Area in the United States (Cochran *et al.*, 1965), were intended to radio-locate a large number of tagged animals at all hours of the day and night, thus providing a great many data for range analysis without the need for extensive mobile tracking. This sort of system is at its best in areas where few topographical errors are likely, especially in flat yet inaccessible terrain, (e.g. wetlands), or where too much disturbance would be caused by mobile tracking (e.g. in very open country or small-mammal habitats).

Three different types of automatic radio location system have been used for tracking wildlife. The Cedar Creek system, which was adopted at Chizé Forest in France (Marquès, 1972; Deat *et al.*, 1980), used rotating twin Yagi antennas at the top of 20–30 m towers about 1 km apart. The Yagi direction was indicated by 360 slits in a drum which rotated past a photocell counter. The rotation speed of such systems must be relatively low, and the tag pulse rates quite fast: if signals occur at 500 ms intervals, an antenna which rotates once per minute will swing through 3° between each signal. Accuracy can be improved by sweeping more slowly (4 min per revolution at Chizé), but this increases the error when each tower scans a moving animal at different times. Another disadvantage is the system's limitation to animals with relatively restricted ranges, because accuracy is low for distant tags: the angle between bearings is then small. If there are only two receiving stations, accuracy is also lost as an animal moves towards the baseline between them. Moreover, some promising high points prove to be disappointing for bearing accuracy, as in mobile tracking. It is important to test this, preferably using tags with a variety of antenna orientations at each test site (Cederlund *et al.*, 1979; Lee *et al.*, 1985). You can then move inaccurate receiving stations, and recognize parts of your study area where fix accuracy is poor.

A second type of direction-finding system, which is widely deployed on boats, might also prove useful for animal tracking. This is based on an "active array" of four dipoles, set vertically at the corners of a horizontal square. The differences in the signal's phase at each dipole are used to determine its direction. Since the determination is effectively instantaneous, it can in theory be made within one pulse. Direction finding is thus less affected by animal movements than is the case with a rotating Yagi, and there are no moving parts to maintain. However, the sensitivity of the system is reduced by the relatively low gain of a dipole, compared with a multi-element Yagi. One system tested for animal location had only about a third the range of a six-element Yagi (T. Parish, personal communication). This may be no problem when receiving powerful navigation beacons, but is a serious drawback for work with the much weaker tags on small animals. Nevertheless, there may be scope for further development of active array systems. The vertical stacking of two or more dipoles at each corner would increase the gain, if this could be done without also reducing the accuracy of the signal phase comparisons.

Among long-distance location systems, the hyperbolic system is most accurate, and technically elegant in its absence of moving parts. A tag's position is defined by differences in the time its signal takes to reach three or more omnidirectional antennas. If the signal takes 10 ns longer to reach one of two stations, the tag is somewhere along a hyperbola which is 300 m closer to one station than the other. The time differences between these stations and a third give two further sets of hyperbolas which intersect at the tag's position (Fig. 7.1). Needless to say, the relative positions of the receiving stations must be known precisely. Moreover, each station must detect the signal at the same point in its development. This makes tag circuits unsuitable, if their pulses do not rise virtually instantaneously to peak intensity. If one receiving station is nearer to the tag than another, and therefore detects the signal closer to its onset (Fig. 7.2, station 3), then the tag's distance to that station will be underestimated. Tags for such a system must transmit powerful signals with a rapid pulse rise-time, and complex coding systems may be necessary so that signals which arrive later than the direct-path signal can be rejected (Yerbury, 1980). In the system at Grimsö in Sweden (Lemnell et al., 1983; Lemnell, in press), the tags are active transponders. On activation by a powerful VHF signal from a master station, each tag sends a coded signal pulse which carries up to three channels of sensor data. Each tag is identifiable, because an integral CMOS clock sensitizes it for activation in its own 2 s slot every 2 min. There are thus 60 tag channels, and four tags can be discriminated on each channel by pulse coding. The VHF tag signals are received at the master station, and at three slave stations which immediately retransmit a UHF signal to the master. Since signal

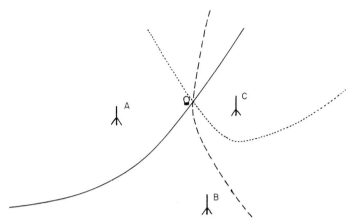

Fig. 7.1. Hyperbolic location of radio tags. The curves represent lines of equal distance, and hence equal signal time, between stations A and B (solid line), B and C (dotted) and A and C (dashed). The tag is at the intersection of the curves.

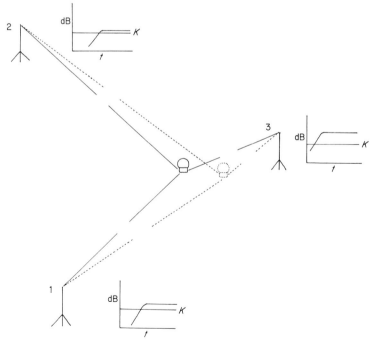

Fig. 7.2. The signal-rise problem in hyperbolic detection systems. Signal strength must reach a threshold (K) to be detected at each station. At station 3, the signal from the tag is detected early, not only because it gets there first, but also because the slow-rising signal is strong and therefore crosses the detection threshold earlier than at the other stations. As a result, the tag position is computed too close to station 3.

processing is done at the master station, and the slaves are only activated when the master transmits, the slave stations can be powered by solar cells and placed 6–10 km apart. A tag signal of 8–20 W is required for detection by several stations over these distances, so "transmission on demand" is important for prolonging tag life. The first tags, weighing 600 g and placed on elk, gave a fix accuracy of about 30 m. Track-tech (see Appendix I) can now supply 40 g tags.

A fourth, much simpler system is suitable for recording the positions of small rodents or other terrestrial animals with small ranges. A set of separate parallel wires is stretched low over the study area in one direction, with another similar set at right angles to them and slightly above them. These wires serve as receiving antennas, the tagged animal's approximate position being given by the intersection of the two crossing wires which receive the strongest signals (Chute *et al.*, 1974; Zinnel and Tester, 1984).

A major drawback with all automatic radio location is that it misses much information which can be gathered by tracking on foot, such as observations of whether the animal is feeding, resting, courting, etc. Moreover, fixed-station recording lacks the flexibility of a mobile tracker to increase accuracy when necessary, for instance to decide which side of a habitat boundary or in which tree an individual is feeding, or indeed whether it is in a tree or on the ground. Although vast quantities of data are collected, many become redundant if statistical independence is to be maintained during the analysis.

B. Manned direction-finding stations

Since automatic radio location systems are relatively expensive to construct and have a number of limitations, relatively few projects have seen fit to invest in them. Manned direction-finding stations have proved somewhat more popular, which may seem surprising in view of the current tendency to automate wherever possible. However, apart from the high cost and risk of commissioning delays with automatic systems, their sensors are typically 6–9 dB m less sensitive than the human ear, which reduces their effective range quite appreciably compared with manned stations. This is particularly important for tracking animals with large ranges or weak tags.

Manned stations can take quite accurate bearings, with null-peak antenna systems (Plate 16), and are easily "programmed" to reject "silly" fixes! If two communicating observers plot each pair of bearings as they are taken, preferably by simply entering bearings into a microcomputer-generated map display (which also stores the data), they can immediately try again if a small mammal seems to have moved 1 km in 5 min. So, for 1–3 year projects in which there is a premium on high sensitivity and accuracy, it is still well worth settling for a pair of stations with manually rotated twin-Yagi (null-peak) antennas. One person can man such systems if the second antenna is remotely servo-driven, and its signals relayed to the second (Smith and Trevor-Deutsch, 1980; Linn and Wilcox, 1982), but beware of the extra time required for development and teething problems.

Beware also of making unreasonable demands on the system. A number of projects have tried unsuccessfully to track foraging seabirds from the shore. For a start, it is difficult to get accurate bearings from Yagis to moving animals, such as birds in flight or bobbing on the water, because changing antenna orientation affects the signal amplitude. Moreover, the bearings from each station become very similar, with resulting large error polygons, when birds are far out to sea. Such tracking might be more successful with active array or hyperbolic systems, using very powerful tags to compensate for VHF absorption across the sea surface. However, it is easy to forget the constraint posed by the horizon. In imperial measures, the horizon distance in miles approximates to the square root of the observer's height above it in

Plate 16. Twin four-element Yagi antennas used for taking null-peak bearings from a hilltop.

feet (in terms of the metric system, multiply distances in metres by 3, and multiply the subsequent root by 1.6). The VHF waves can be detected slightly over the horizon, by a factor of up to 1.4 in excellent propagation conditions, but more often by about 1.2. This means that a Yagi on a 100 m headland will probably not detect signals much more than 33 km out to sea, or perhaps 50 km for a bird flying at 20 m ASL, even if the tag could be heard at 200 km in line-of-sight from an aircraft.

II. PRESENCE/ABSENCE RECORDING

The most frequent use of automatic stations has been to record when radio-tagged animals visit a particular site. For example, if a Yagi antenna is pointed vertically at a tree nest, and coupled to a receiver with a chart recorder, a steady recording peak will indicate the animal's presence in the nest, whereas a trace on the base line indicates that the animal is out of range. Rustrak chart recorders (see Appendix I) have long been used for this purpose (Williams and Williams, 1970; Gilmer et al., 1971). These analogue recorders also show an intermediate, irregular trace when the animal is active near the nest (Plate 17). To provide a steady trace from the pulsed radio signal, a large electrolytic capacitor (e.g. 2200 μF, RS 105–313) is put in parallel with the output from the receiver's recorder socket. One of the earliest presence/absence systems was for seals visiting ice-flows (Siniff et al., 1969), and riverside systems are now used to record the passage of tagged fish, or when they leave a site to run upstream (Solomon and Storeton-West, 1983).

Whereas a simple tracking receiver can only be used to record one tag frequency at a time, programmable receivers can be used to record visits by tagged individuals on several different frequencies, for instance to a breeding colony. Whenever the receiver samples a new frequency, a custom-built electronic controller resets the baseline progressively across a wide chart, returning the marker to its true baseline at one side of the paper at the end of each sample cycle (J. French, personal communication). Each step between tag frequencies literally creates a step on the record, with a signal being marked on that step if it was present on that particular frequency (Fig. 7.3). Disadvantages of this system are the expense of large chart recorders, and the need for a rapid chart speed to minimize the sampling interval for each frequency. If the marker must move 3 mm for signals to stand out clearly on each step, and each frequency must be scanned once every 5 min, the chart speed must be 180 mm h^{-1} (4.32 m day^{-1}) for every five tag frequencies scanned. Moreover, Rustrak chart speeds often do not remain at the nominal value.

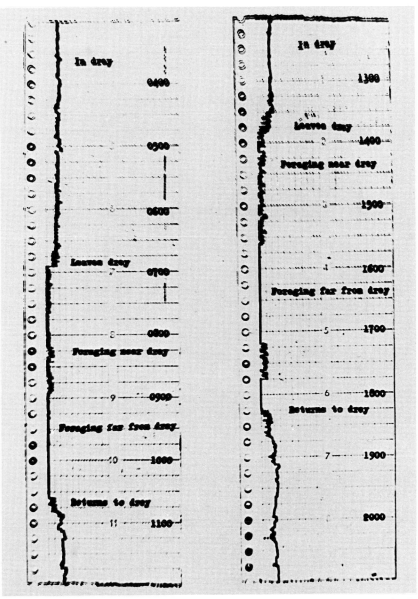

Plate 17. A chart obtained on a Rustrak recorder, using a Yagi antenna pointed at a squirrel's nest. The signal was strong and steady when the squirrel was in the nest with its young, irregular when it was foraging nearby, and the trace was on the baseline when the animal was at least 50 m away.

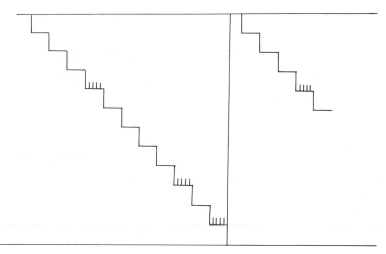

Fig. 7.3. A multi-channel chart from a recorder which stepped each time the receiver frequency was changed. Signals were detected on channels 4, 9 and 11.

If a single tagged animal is likely to be present or absent for long periods at a time, intermittent sampling can be used to save the expense of buying a chart recorder, and to conserve the system's power supply. In this case, an inexpensive tape recorder can be used to record the receiver's audio output. You can build a mechanical timer to switch on the receiver and recorder (Downhower and Pauley, 1970; Göransson, 1980), but an electronic timer is more reliable and flexible (Macdonald and Amlaner, 1980). Using 10 s samples at 10 min intervals, a 24 h record can be taken on a 45 min cassette (60 min cassettes are less reliable on cheap recorders). Since high-fidelity recording is unnecessary, tape recorders can sometimes be modified to run at one-half to one-fifth the normal speed. Sampling can then be more frequent, or tapes changed less frequently, and the record analysed at normal speed. A major disadvantage with this inexpensive system is the difficulty of timing events. The tape footage counter provides an estimate of the time since the start of recording, provided that the tape moves at a constant speed through the recorder, but this condition is often not met. A more accurate but laborious method is to count the number of sample periods while analysing the tape. There may be a characteristic noise recorded each time the recorder shuts down; otherwise you can arrange for the timer to emit a bleep or loud click at the beginning or end of each sample.

A multichannel system has been developed at the Ministry of Agriculture, Fisheries and Food Laboratory in Lowestoft, based on a Yaesu programmable receiver linked by a microprocessor to a cassette data recorder and a small thermal printer (Pearson, 1986). Each of the 10 Yaesu memory channels is scanned every 5 min. If the riverside system detects that a radio-tagged salmonid has migrated into a nearby lie, it prints the channel number and time, and makes a 15 s tape recording for further verification. If a tag is detected for more than 12 consecutive scanning periods, printout and tape records are curtailed until there is no signal detection, at which point the tag's departure is logged. There is an additional printout of signal presence or absence every midnight, to show that the system is working, and a range of environmental variables can be recorded with appropriate sensors. The system is enclosed in a weatherproof box, and runs for about 7 days on internal batteries, or 70 days with an external car battery.

Printers, chart recorders and tape recorder systems, with their moving parts which may corrode in the damp and slow or even stop in cold weather, will most probably soon be replaced by solid-state RAM systems, such as the Grant "Squirrel" (see Section V), or at least used only as back-up for RAMs. Relatively little memory is required if the system can be programmed merely to store the time once when a signal is acquired, and once when it is lost.

III. RADIO TELEMETRY

Pulse-modulation telemetry has been used in many wildlife studies, especially for temperature recording. Tags with thermistors have been used to study temperature regulation in several reptile species. Some studies have measured deep body temperature with ingested tags (Mackay, 1964; Osgood, 1970; Swingland and Frazier, 1980), with implanted tags or with rectal probes from external tags (McGinnis, 1967). Other projects have used thermistors mounted on external tags to record ambient temperature in the animal's chosen microhabitat (Standora, 1977; Standora et al., 1984).

To detect signals from an animal throughout its range, a high-gain receiving antenna is needed. To cover a large area, the antenna should be mounted on a high point and preferably at least 10 m above the ground, but with the cable to the receiver as short as possible to reduce signal losses along the wire. If the antenna is to be near the centre of the study area, use an omnidirectional type. Many projects use $\lambda/2$ dipoles with ground-plane elements (Amlaner, 1980). However, a higher gain, equivalent to a Yagi with three or four elements, can be obtained with three or four folded dipoles in a vertical stack, provided the cables from each are exactly the same length, or with a $5\lambda/8$ over $5\lambda/8$ over $5\lambda/8$ collinear antenna. Alternatively, you may decide that the receiving system is best positioned to one side of the study area, and then use a three- or four-element Yagi to receive the signals.

When siting a valuable recording station, security is often as important a consideration as optimizing the signal reception. If there is no building in which to lock the receiver and logger, it may be best to bury them under the ground in a suitable container (e.g. a plastic dustbin), taking care also to hide or camouflage the antenna and its cable. As a last-ditch deterrent, some projects put "RADIOACTIVE", "BIOHAZARD" or "HIGH TENSION ELECTRICITY" warning labels on and in the equipment's container!

If only one animal is to be monitored at a time, signals can be recorded directly with a tape recorder, and the signal pulses later counted for each sampling period. This provides a cheaper system than some form of electronic pulse-counting or pulse-interval measurement, and is also better for detecting very weak signals. However, counting pulses is not as accurate as electronic pulse-interval timing, and further minor inaccuracies can arise from changes in the tape speed. A tape recording system is at its best for recording large changes in pulse-intervals, from posture sensing tags, or from thermistors whose cooling is used to indicate flight or foraging excursions from a warm nest.

If many frequencies are to be monitored at a time, an electronic signal-processing system must be used. If each received pulse can be used to trigger a regulated voltage (e.g. 5 V) spike, the pulses can be counted over a predetermined sampling period by several makes of commercial data logger. The simplest of such systems can be based on a Grant Squirrel data logger (see Appendix I), which records data in an 8 K–32 K RAM for later output to a computer with an RS232 interface. One version of the Squirrel will count pulses for a preset sampling period, at preset intervals, and store each count in one byte of memory. In a typical single-channel recording system, with a sampling period once every five minutes, only 288 bytes of memory would be filled each day. The total cost of recorder, software and accessories is about £1000.

This logger can also be used for multichannel recording, if the sampling periods are triggered by a high-accuracy external timer which simultaneously changes the channel on a programmable receiver. With the receiver set to scan 20 channels, the first channel's signals are counted again after the count for all 20 channels has been stored. A simple analysis program will then list each count in one of 20 columns, with the time for the start of each 20-channel scan. With the sampling period set at 15 s, continuous sampling will monitor each channel once in every five minutes, and the system will fill 5.76 K of memory each day. A 32 K Squirrel in such a system must have its memory dumped every 5 days. This is a convenient interval at which to check automatic stations, if you do not want to risk losing very large quantities of data when something goes wrong!

A few projects have built more sophisticated, microprocessor-controlled logging systems for telemetry data. A logger made in the Institute of Terrestrial Ecology sets the system running only when animals are likely to be active (e.g. during daylight), controls the length of sampling periods by giving the programmable receiver a cue to change channels, and records the interval between each pair of pulses during each sampling period. If it is monitoring signals with a relatively constant pulse interval, from tags with thermistors rather than posture sensors, a controlling EPROM can be programmed to reject any readings which differ by more than a chosen percentage from the overall average, and then to store the new average figure. This reduces problems arising from interference, or from failure to register some pulses in a sequence. The reading is shown on a liquid crystal display to check that the system is operating properly, and is stored, with the channel number and time at the start of each scan, in a 1 K RAM which is dumped regularly on to a data cassette. Several hundred kilobytes of data can thus be stored in a form which is secure against interruptions in the power supply. Since each reading is uniquely labelled, memory can be saved by not storing zero readings (e.g. from faint signals), and the data file is not fatally corrupted if there are occasional errors when the RAM is output to tape.

At least two logging systems for radio-telemetered data have been built round microcomputers (Howey *et al.*, 1984; Cooper and Charles-Dominique, 1985). There is plenty of scope for exploitation of portable microcomputers, with integral cassette recorders or large battery-backed RAMS, as inexpensive "intelligent" data loggers. For instance, they could be used for sophisticated mortality sensing. Such a system would use tags with simple activity switches, without the special mortality-detection circuitry. Instead, the computer could be programmed to trigger an alarm if the signal pulse rate did not change on each channel over a preset period, which could even be confined to times of day when the animal is normally active (to avoid "false alarms" during normal rest periods). Such a system could also be programmed to sound an alarm upon prolonged loss of a signal, due perhaps to radio failure or dispersal of a tagged animal, not to mention the "mundane" opportunity for logging activity data.

There are of course many animals which spend much of their time beyond the range of any fixed station. A recent development for such cases has been to build a data logging capability into the tag itself, with the tag's memory either being read when the animal is recaptured, or transmitting the data at high speed to a satellite or surface-based receiving station when appropriate (see Chapter 2.V.F). Such systems have been used to monitor diving parameters for marine animals, but their cost is likely to remain beyond the reach of most biologists.

8
Some Analysis Techniques

A major aim of most wildlife radio tagging is to collect behavioural or demographic data. This last chapter takes a look at different ways of handling these data. Its final section introduces a new range analysis technique, which is listed as a BBC BASIC program and described more fully in Appendix III.

I. DENSITY ESTIMATES

To make density estimates, radio tags have been used both as Lincoln-Seber Index markers and as a means of correcting for grid edge effects. For instance, the number of goshawks visiting study areas in Sweden was estimated by checking whether each hawk sighted there was radio-tagged (Kenward *et al.*, 1981). This gave a number (r_i) of tagged hawk observations among s_i sightings, which were counted only if they were made "by chance" during routine travel about the study area, and not while a particular hawk was being tracked. There were new hawks marked and ones which emigrated throughout the study seasons, but the number of radio-tagged birds available for sightings (R_i) was known in each period (i), and the total number (N_i) during each period area could be estimated by:

$$\frac{(R_i + 1)(s_i + 1)}{(r_i + 1)} - 1$$

with variance:

$$\frac{(R_i + 1)(s_i + 1)(R_i + r_i)(s_i + r_i)}{(r_i + 1)^2(r_i + 2)}$$

The overall estimate N, which was the mean of the N_i estimates weighted by the inverse of their variances, did not represent the density of hawks in each area, because the birds were also foraging away from the study areas. There were in fact two density estimates. One (d_t) was derived from the proportion of time that each radio-tagged hawk spent in the area, and the other (d_a) was based on the proportion of each hawk's range that was in the study area. If the time-based density was larger than the area-based density, then the study area was a favoured part of that hawk's range.

An assumption behind these estimates was that tagged and untagged hawks were equally likely to be sighted. In one of the study areas there was evidence to justify this assumption, inasmuch as a kill made in the area represented time spent there. Among the many pheasant kills found by game wardens who were not involved in the tracking, the proportion which had been made by tagged hawks was the same as the proportion of tagged hawks in sightings (Kenward, 1977).

This Lincoln–Seber Index estimation is suitable for elusive species, which are often sighted too briefly for a visual marker to be spotted. The individuals must of course be widely dispersed, or the signal coming from a sighting position might be coming from a tagged animal near to the untagged one that was seen. At the other end of the dispersion scale, however, for animals which live densely enough for their abundance to be estimated by grid trapping, radio tagging can be useful for deriving more accurate abundance estimates than would be possible from the trapping alone. This is because of problems in estimating densities from grid trapping if the grid does not cover all the habitat used locally by the species. The grid may cover only part of a wood or a grassland, so that some of the animals caught on the grid normally spend little of their time on it, and may even have been drawn onto it only because of the baited traps. In this case the density would be overestimated if the number of animals caught was simply divided by the grid area— assuming that all the animals on the grid had been caught in the first place!

There are at least three ways of using radio tagging to improve the accuracy of density estimates from grid trapping. If ranges are relatively compact (not elongated), and one assumes that the distribution of ranges is even across the edge of the grid, the area considered to be covered by the grid may be extended at the edge by half the average radio-range radius (Kenward, 1985).

A more sophisticated approach is to relate the proportion of a tagged animal's range or time within the grid to the distribution of traps in which it was captured (Tonkin, 1983). The best-fit function may be the result of a simple or a multiple regression, perhaps involving the number of different traps in which each individual was captured, the number or proportion of these traps at the edge of the grid, or the number of captures per trap in each category. This function is then used to estimate the proportion of each untagged animal's time (or range area) spent on the grid, assuming that tagged and untagged animals used the grid area in the same way. This is a reasonable assumption if tagging was unbiased: for instance by tagging every nth individual at first capture.

A simpler approach, if the tagging was unbiased, is to estimate the grid area density as (the number of animals caught, or estimated to be present) × (the proportion of their time, or range area, that tagged animals spent on the grid). A neat trap grid is not necessary for this third density estimate, whereas for the first two approaches the grid spacing must be small enough to put several different traps within each animal's range. On the other hand, an advantage of the first estimate is that the edge correction may be applied to areas in which there is no radio-tagging, if the trap-based ranges are similar in each area.

II. SURVIVAL ESTIMATES

When radio-tagging is used to estimate survival and dispersal rates, the calculations are often complicated by animals entering a tagged population over a number of months, instead of fledging or weaning over a relatively short period. Moreover, animals leave the tagged population through tag extinction (including component failures and cell exhaustion) as well as through death or emigration. The best approach is therefore to estimate the rates on a monthly, weekly or daily basis: the size of interval chosen will depend on how accurately you can assign the deaths. Animals whose signals are lost are included in the population for every interval prior to the one in which they disappeared, but are omitted altogether from that interval. By analogy with Trent and Rongstad (1974), the survival rate for each interval (i) is then:

$$\frac{\text{number of tags working during interval } (n_i) - \text{number of deaths in } i \text{ interval } (d_i)}{n_i}$$

It is important to note that survival estimates will be biased upwards by any failure to distinguish "routine" tag extinctions from other sources of signal loss. This is a particular problem with animals which can only carry a weak tag, if they disperse frequently. Unless you check them often enough to record their departure from the study area, you may never detect the signal again, and assume that the radio has stopped transmitting.

Even if dispersal can be ruled out by efficient searching (Chapter 6), survival will be underestimated if tags are destroyed by traumatic events which also kill the animal (e.g. predation, road accidents, shooting). However, provided that you kept a regular check on tag pulse rates, and have access to subsequent trapping or sighting data, you will be able to set a lower limit to your survival estimates. To do this you should include as deaths all tagged individuals whose signals were lost (a) well before the end of their expected cell life, (b) with no slowing or irregularity in their signals and (c) with no subsequent recapture or resighting.

This estimate will be biased downwards by inclusion of undetected emigrants, and animals which survived a "symptomless" tag failure, but emigrated or died before they could be trapped or seen again. However, the frequency of such failures can be estimated, because you should have recorded similar loss of signals from some of the animals which you did see or trap again. The best survival estimate (s_i) for each interval includes the number of animals actually found dead (d_i), plus the number of tags which disappeared unaccountably (u_i), minus the tag failure rate (c_i):

$$\frac{n_i - d_i - u_i}{n_i} + C_i$$

If intervals are relatively long and contain many deaths, binomial confidence limits can be fitted to individual s_is to compare rates between different intervals. However, greater precision is obtained if the standard error can be estimated for a number of daily rates (S_i) within a span. If some intervals within the span are L days long, the survivals can be converted to daily rates since:

$$s_i = S_i^L$$

The S_i distribution may well be skewed, but can probably be normalized with a natural log transformation (Bart and Robson, 1982; Heisey and Fuller, 1985). The total survival for the time span (S^*) is then the product of the S_i^L's within it. Confidence limits are obtained by multiplying the SE (standard error) by the appropriate value of Student's t for the sample size, working in logarithms if a log transformation has been applied.

The results for each interval may also be expressed as a mortality (i.e. 1 − survival rate). Since radio tagging can be used not only to record deaths, but also to provide relatively unbiased data on different causes of death, you may well want to estimate partial mortalities. If a factor causes d_{ij} out of a total d_i during an interval in which survival is S_i, its partial mortality rate (M_{ij}) is:

$$(d_{ij}/d_i)(1 - S_i)$$

However, this relationship does not give the best partial mortality estimate for a span, such as a year, in which data are available for a number of intervals: two factors might cause equal numbers of deaths during the year, but the factor which causes most deaths at the end of the year, when the population size is smaller, will have the most influence on total mortality. Heisey and Fuller (1985) show that the partial span mortality (M_j^*) can be estimated by the sum of the partial mortalities, for each interval, multiplied by the product of the survivals for all preceding intervals:

8. Some Analysis Techniques

$$\sum_{i-1}^{1} M_{ij} \prod_{k=0}^{i-1} S_k$$

where $S_0 = 1$. Confidence limits may be estimated from daily mortality rates in each interval, as for the survival rates.

You may also wish to study relationships between survival and other variables, such as habitat type, experimental treatment, or the age and weight of each animal. If animals are tagged within a fairly short time, and reasonable numbers fall into "alive", "dead" and "unknown" categories at a later date, the effect of the variables during the intervening period can be investigated with a discriminant function analysis (e.g. Snedecor and Cochran, 1971). An alternative approach, which is more suitable for animals marked over a long period, but which requires a binary classification of the independent variables (e.g. male/female, large/small, cover/no cover), is the Cox proportional-hazards model (Cox, 1972; Kalbfleisch and Prentice, 1980). This multiple-regression model, in which the dependent variable is the probability of dying (the hazard rate) at a particular time, has been used to analyse snowshoe hare data by Sievert and Keith (1985).

III. EVENT RECORDS

A major advantage of telemetry over techniques such as trapping, or "chance" observations, is that it records events in relation to time. Instead of using captures in traps to register the places visited by an animal (with the animal's foraging perhaps biased by the traps), radio tracking can show not only how often an individual visited each part of its range, but also what time of day it went there. Instead of merely describing the diet of a predator species (with possible observation bias if this involved searching for prey remains), radio surveillance can reveal the kill-rate for each prey by each individual predator.

The data from radio surveillance, telemetric monitoring and presence/absence recording tend to be continous sequences of events along a time-base. They are most often summarized as event frequencies (e.g. beats per minute, visits per hour, kills per day), as durations between events (time submerged, time since sunrise) or as proportions of activity periods (percentage of time spent grooming, in woodland, with mother). Provided that reasonably independent samples have been taken, non-parametric statistical analyses may be used to compare the resulting measures, or more sophisticated analyses used to look for associated sequences (runs) of events. For example, Roth (1983) describes an iterative chi-squared procedure which identifies the main changes in activity, measured from hour to hour throughout the day. As discussed in Chapter 6, it is generally better if the data were collected by following, say, eight animals for an hour each, than from eight hours of following one animal.

IV. DISTANCE AND SPEED RECORDS

In estimating energetic costs, the duration of locomotion or the time outside a warm nest may be measured by telemetry, if tags have suitable sensors. Where this is impractical, however, you can use the distance moved as an index of energy expenditure. During radio surveillance, for instance, every perch position can be recorded, or a fix taken at 5 min intervals, to estimate the total distance moved each hour, or on each foray from a nest. During systematic range recording, there may be no more than three fixes registered at set times each day, with a roost site. Nevertheless, the daily sum of the four distances between consecutive fixes (e.g. roost–midmorning, midmorning–midday, midday–afternoon, afternoon–roost) can give a useful (albeit very approximate) index for comparing the movement of different individuals over a number of days.

If you are monitoring individual movements continuously, and wish to compare speeds (rates of change of position) between different animals or different times of day, you will probably have measurements over a variety of time intervals (e.g. 210 m in 7 min, 495 m in 4 min, 125 m in 11 min, etc.). These can be converted into distances per minute, and compared by using Mann–Whitney U tests or appropriate non-parametric analyses of variance (e.g. Siegel, 1957).

V. RANGE ANALYSIS

If an animal's position is sampled at intervals over a period of time, a plot of the fixes indicates the animals home range for that time. If the fixes are taken at fairly regular intervals, they also sample the animal's distribution in time through that range. From this concept, of sampling the animal's position along a time-base, different approaches to range analysis have evolved.

A. Grid cells

One approach concentrates on the grid cell in which each fix occurs (Siniff and Tester, 1965). Cells may be hexagons, to equalize distances between their centres, but it is often easier to use grid squares, whose sides should not be smaller than the resolution of the tracking technique. In its simplest form, this approach is good for analysing habitat use. To represent the time an animal spent in each habitat type, you can count the number of cells with fixes in each habitat, and express this as a proportion of the total number of cells with fixes. Habitat choice is shown by comparing these proportions with the proportion of all the cells on the map which contain that habitat type (e.g. with a chi-squared test). If you are satisfied that multiple visits to the same square are statistically independent, for instance because the animal was recorded in distant squares between each visit, you may emphasize the choice by multiplying the number of cells in each habitat by the number of fixes in each, as a proportion of the total number of fixes.

The grid cell approach is also good for investigating interactions between different individuals (Adams and Davis, 1967). The proportion of each animal's grid cells which are visited by another gives one measure of their "static" interaction. Alternatively, as in habitat analysis, the intensity of overlap may be emphasized, by using (overlap cells) × (average of fixes in each) as a proportion of the total fixes for each animal. Provided that one animal's fix was always recorded shortly before or after the other's, you can also analyse their association in time, or "dynamic" interaction (Macdonald et al., 1980). If the distribution of distances between each contemporary pair of fixes is significantly smaller than a distribution generated by pairing their fixes at random, then those individuals tended to associate. Tests for the significance of dynamic interactions are given by Dunn (1979) and Macdonald et al. (1980).

If there are enough fixes, so that several might occur in any one cell, you can calculate indices of variation in each animal's spatial use of its range. Ecological diversity indices may be used, or the coefficient of variation (standard deviation divided by the mean), but these only show whether the fixes tend to be clumped, and not whether they form one or many clumps. Macdonald et al. (1980) pointed out that the Rasmussen Index (Rasmussen and Rasmussen, 1979) overcomes this, by taking the distance between cells into account: this index reaches its maximum value when intensively used cells are close together. A value (d) is calculated for each of the P (= $N(N-1)/2$) cell pairs, as the sum of their fixes divided by the distance between them, and used to derive the index

$$\frac{\Sigma(d_i - \overline{d})^2/P}{\overline{d}}$$

The grid cell method is at a disadvantage, compared with other range analysis techniques, when comparing range areas. Measuring the range area

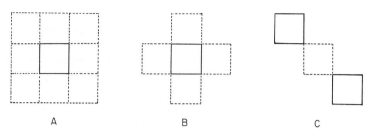

Fig. 8.1. Some rules for increasing the area of grid cell ranges. In the "queen's case" (*A*), all cells adjacent to a used cell are considered to be used. In the "rook's case" (*B*), only the cells which share a side with a used cell are added to the total. "Linked cells" (*C*) are added if they lie between cells which were used consecutively.

as the product of each cell's area and the number of cells which contain a fix, it takes far more fixes to reach sampling saturation (Chapter 6), by "filling-in" all the squares that the animal is using, than with other methods of estimating range area (Voight and Tinline, 1980). This problem can be alleviated, albeit at the expense of accuracy, by using a variety of rules to include cells beside or between those which contain fixes. For instance, the "queen's case" adds all eight adjacent ("influence") cells, the "rook's case" adds only the four whose edges join, and a "linked cell" method adds unfilled cells between pairs of fixes, on the assumption that the animal visited these intervening cells on its way between the sampling points (Fig. 8.1).

B. Outline techniques

The simplest way to assess range area and position is to draw the smallest possible convex polygon round the outermost trap sites or radio fixes (Dalke and Sime, 1938; Mohr, 1947; Southwood, 1966). Although range analyses have often been based on minimum convex polygons, which can be overlaid on other ranges and habitat maps to give an indication of habitat use and static interactions, they give no indication of how intensively the animal uses different parts of its range, and the area is strongly influenced by peripheral fixes. Since many species seem to make occasional excursions well outside their normal foraging range, minimum convex polygons often contain large areas which animals never visited, and are sometimes more representative of areas traversed during excursions than of where an animal spent most of its time (Fig. 8.2).

Several techniques have been proposed for reducing the unvisited areas. The range outlines can be made concave (Stickel, 1954), perhaps by drawing the peripheral line to an "internal" point wherever the distance between

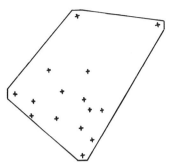

Fig. 8.2. A minimum-area polygon. Two outlying fixes are responsible for adding as much extra area as is enclosed by all the other fixes.

edge points is greater than a quarter of the range width (Harvey and Barbour, 1965), or by joining the fixes in order of distance from the arithmetic centre of activity (Voight and Tinline, 1980). Another approach is to exclude areas of unused habitat (Ables, 1969). There is an arbitrary element to these approaches, however, which makes them somewhat unsatisfactory.

(1) Parametric range models

Hayne (1949) suggested that an animal's (trap) range could be seen as a centre of activity, defined as the mean of the x and y coordinates, surrounded by zones of decreasing probability of trapping the animal. A way of calculating this probability distribution was provided by Dice and Clark (1953), leading to the description of ranges as probabilistic circles (Calhoun and Casby, 1958; Harrison, 1958), and subsequently as ellipses (Jennrich and Turner, 1969; Mazurkiewicz, 1971; Koeppel *et al.*, 1975; 1977). These range models assume that the points are independent and distributed normally about the activity centre. Although fixes obtained by continuous tracking are not independent, Dunn and Gipson (1977) provided an elliptical model which takes this into account. Alternatively, the data set can be sampled, using an autocorrelative technique to estimate the minimum time interval for independence (Swihart and Slade, 1985).

The value of these models is in defining a core range, in which an animal spent most of its time, by excluding peripheral locations. In the case of radio fixes which have been sampled at regular intervals, such that each fix is a sample of the animal's position along a time-base, exclusion of the area represented by the outermost 10% of fixes gives a 90% probability that the animal was in the rest of its range at any time during range recording. However, there is a serious problem with these circular and elliptical range models: fixes are often not distributed normally about an activity centre, and the resulting circles or ellipses often contain large areas into which an animal never ventured (Adams and Davis, 1967; Macdonald et al., 1980). Although probabilistic circles and ellipses have some value for comparing range sizes, they seldom fit the fixes well enough to be used for rigorous comparisons of range overlaps or habitat use.

A much better fit can be obtained with models that allow for clumping in the data. Don and Reynolls (1983) have developed a sophisticated parametric multinuclear model, which assumes that fixes may be distributed about several activity centres in a range. In many cases this will produce probability contours with a reasonable fit to the data, the exception being when many fixes lie on the edge of the range. In this case the local distribution is highly skewed, and too many fixes occur in "low probability" areas.

(2) The harmonic mean model

Following work by Neft (1966), Dixon and Chapman (1980) showed how the first inverse moment of the fixes could be used to estimate an activity centre, and as an elegant way of drawing range-use contours (isolines). The first inverse moment of the fixes is calculated for each grid intersection, by summing the inverse of the distance to each fix (Fig. 8.3A), dividing by the number of fixes, and then inverting this value. The harmonic mean centre is estimated by the grid intersection where the value is minimal. Since the most distant fixes have the lowest inverse moments, they have relatively less influence on the harmonic mean centre than when they are used to calculate an arithmetic activity centre position. The harmonic mean is therefore the more satisfactory range centre estimate. Its drawback is the high inverse moment contribution from fixes which are very close to grid intersections. The correction is to give fixes in adjacent grid squares a standard contribution to the moment ($=1$). Nevertheless, such fixes make the estimated centre very dependent on the position and spacing of the grid: changing the grid can alter the position quite considerably. Spencer and Barrett (1984) therefore suggested that the mean moment estimates be based on the fixes themselves, rather than on the grid intersections. The harmonic mean centre is then a unique position: the fix with the minimum mean inverse moment. These authors also give indices of deviation and kurtosis for the fix distribution. Their unique centre is also ideal for investigations of range spacing, for instance by using nearest-neighbour analysis (Clarke and Evans, 1954).

The value calculated for each grid intersection represents the harmonic mean distance of fixes at that point, and provides a matrix of values on which to draw distance isolines. The distorting effect of fixes close to grid intersections can be reduced by adjusting the fix coordinates to the centre of each

8. Some Analysis Techniques   173

grid square (Spencer and Barrett, 1984). An average distance value for the four corners is then attributed to the centre of each square, and isolines for a particular distance can be interpolated between values which lie on either side of them (Fig. 8.3B).

(3) Core convex polygons

Another non-parametric approach is to base the probability contours on the points themselves (Mohr and Stumpf, 1966). A number of rules can be used

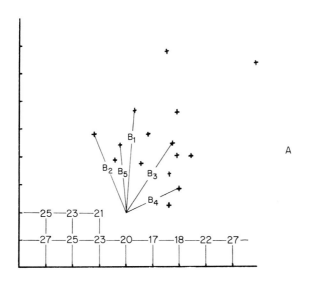

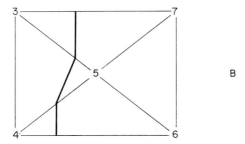

Fig. 8.3. The computation technique for harmonic mean contouring. The inverse mean distance from all the fixes is estimated at each grid intersection. In A, computation has reached the fourth intersection in the second row, and summed the inverse distance to the first five fixes from that point. In B, a central value for each grid square has been estimated as the mean value of its corners, and the isoline with a value of 4.5 interpolated on the edges and diagonals.

to produce "probability polygons" in this way. Simple rules exclude the furthest fixes from a nest, or from the median x and y coordinates, or from the arithmetic mean or harmonic mean of the coordinates. A more complex rule recalculates the arithmetic activity centre after each point has been excluded, and thus focuses on the area with the densest fixes.

The operation of these rules is demonstrated by Program 1, in Appendix III. The program is in BBC BASIC, but could be modified to run in other BASIC dialects with adequate graphics commands (e.g. Sinclair BASIC dialects, or Locomotive BASIC 2). Figure 8.4 shows core polygons obtained by excluding the furthest fixes from a squirrel's nest. Note that the polygon outlines include a "boundary strip" at their edges. This strip allows for each fix to be anywhere within a cell whose sides equal the resolution of the tracking method.

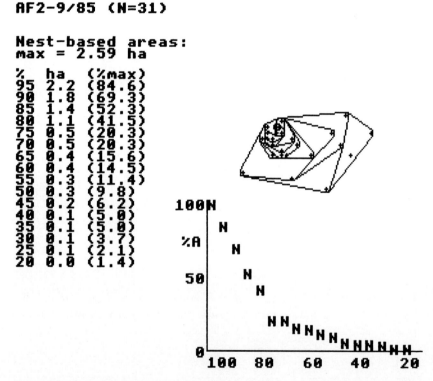

Fig. 8.4. A polygon plot and utilization distribution produced for squirrel data by Program 1 (Appendix III). Polygons reduce in size as the furthest fixes from the nest are progressively excluded. The area decrease is most rapid as the first 25% of the fixes are excluded, so the innermost 75% may be used as the range core.

8. Some Analysis Techniques

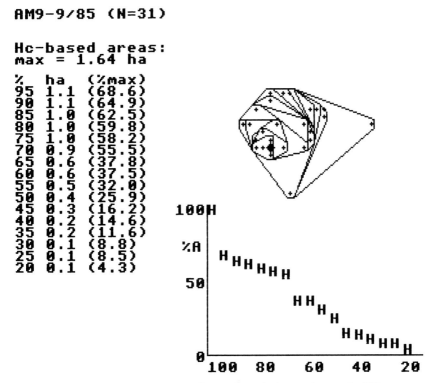

```
AM9-9/85 (N=31)

Hc-based areas:
max = 1.64 ha
%   ha    (%max)
95  1.1   (68.6)
90  1.1   (64.9)
85  1.0   (62.5)
80  1.0   (59.8)
75  1.0   (58.2)
70  0.9   (55.5)
65  0.6   (37.8)
60  0.6   (37.5)
55  0.5   (32.0)
50  0.4   (25.9)
45  0.3   (16.2)
40  0.2   (14.6)
35  0.2   (11.6)
30  0.1   (8.8)
25  0.1   (8.5)
20  0.1   (4.3)
```

Fig. 8.5. A utilization analysis based on distance from the harmonic mean fix. This mononuclear analysis produces a step in the plot, at the 70% level, when all the fixes in the furthest range nucleus have been excluded.

The display from Program 1 (Fig. 8.4) includes a plot of polygon area against the proportion of fixes that are included, called a "utilization distribution" by Ford and Krumme (1979). The polygon area drops sharply as the first few outlying fixes are excluded, and then declines less steeply. This discontinuity (at 80% in the figure) provides a method of defining the animal's core range.

Fix exclusion from a range centre is convenient for fitting probability polygons to mononuclear ranges. However, this approach progressively excludes the smallest nuclei in multinuclear ranges, which results in a "stepped" utilization distribution (Fig. 8.5).

(4) Cluster analysis

Multinuclear probability polygons can be derived by two-dimensional cluster analysis. Program 2 in Appendix III uses hierarchical incremental cluster analysis, with a "nearest-neighbour" joining rule. This rule causes clusters to chain along line features (e.g. hedges, rivers), which seems to fit data better than if the average distance from all members of a cluster is used to decide whether a fix joins. The analysis starts by finding the three fixes with the smallest mean distance between them. This first nucleus has a minimum nearest-neighbour distance to the next nearest fix. The next potential cluster is based in the same way on a fix outside the first cluster, but is started only if the mean distance within it is less than the distance to the first cluster's nearest neighbour (Fig. 8.6). If the nearest neighbour to be added is already assigned to another cluster, the two clusters merge.

When each 5% increment of the fixes has been assigned, the total multinuclear area is estimated as the sum of the polygon areas enclosing each cluster. In an essentially mononuclear range there may be several core nuclei, but these will eventually merge to a single polygon when most of the fixes are included. If this merging occurs only when the last 5% of fixes are

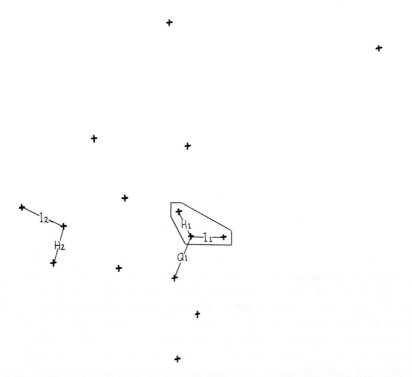

Fig. 8.6. The cluster analysis mechanism. The first cluster of three fixes has the smallest distance $H_1 + I_1$. The distance to their nearest neighbour is Q_1. The next smallest cluster of three fixes has a nearest neighbour distance, I_2, from the two closest fixes to the third fix. This cluster will form next if I_2 is less than Q_1. If not, cluster 1 will gain its fourth fix.

AF2-9/85 (N=30)

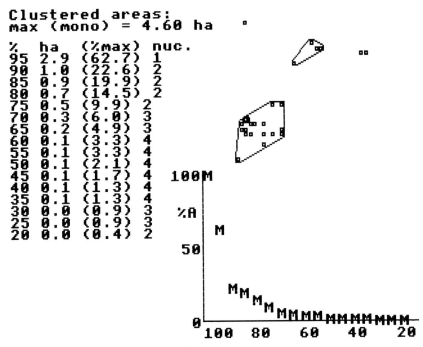

Fig. 8.7. A utilization distribution by cluster analysis, from Program 2. The core is at 90%, and polygons are plotted round the two resulting clusters.

added, because these outlying fixes "force" large core clusters to merge (Fig. 8.7), the range may be considered multinuclear.

The main advantage of the clustering technique over harmonic mean contouring is that the clusters of fixes are treated separately, whereas each value in the isoline matrix is calculated from all the fixes. Isolines may therefore fail to match the fixes accurately, for instance through being "pulled" towards large numbers of distant points, or by contouring round large unused areas when most of the fixes have been included. The latter effect shows very clearly in Fig. 8.8, which was produced by Program 3 (Appendix III) for the data shown in Fig. 8.7. A similar problem occurs when there are multiple fixes in one place, for instance in a favourite food tree: the isoline draws round the site at a distance which depends on the number of fixes, and may thus include quite a large area of unused habitat. With a boundary strip, cluster analysis attributes an area of one "resolution cell" to such a site.

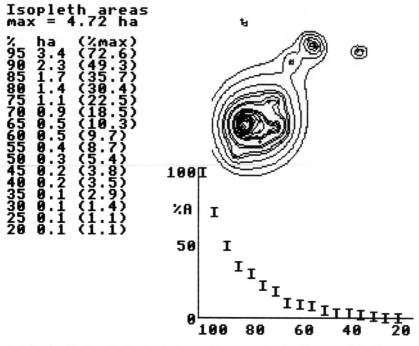

Fig. 8.8. A utilization distribution by harmonic mean contouring (Program 3) for the range data shown in Fig. 8.7. The different analysis shows the same major range nuclei, but isolines beyond the 75% contour do not match the fixes well, and the utilization distribution shows no distinct range core.

The clustering technique is, in effect, an amalgamation of grid cell and outline techniques. It tends to be more accurate than harmonic mean contouring, especially when there are many outlying fixes. For instance, a preliminary run through two sets of 27–29 squirrel ranges has shown that distinct cores are most easily determined from utilization distributions which use cluster analysis, and that these cores are less variable in size than those derived by harmonic mean analysis. On the other hand, harmonic mean analysis produces the best range centre and fix distribution estimates, and provides the most elegant way of displaying range data. The ultimate test, however, is for the technique which gives the most significant results, for example in analyses of habitat use and territoriality. After all, radio tagging and subsequent range analysis are not an end in themselves, but tools for the interpretation of biological phenomena.

References

Ables, E.D. (1969). Home-range studies of red foxes (*Vulpes vulpes*). *Journal of Mammalogy* **50**, 108–120.
Adams, L. and Davis, S.D. (1967). The internal anatomy of home range. *Journal of Mammalogy* **48**, 529–536.
Airoldi, J.-P. (1979). Etude du rythme d'activité du campagnol terrestre, *Arvicola terrestris scherman* Shaw. *Mammalia* **43**, 25–52.
Anderka, F.W. (1980). Modulators for miniature tracking transmitters. In "A Handbook on Biotelemetry and Radio Tracking" (C.J. Amlaner and D.W. Macdonald, eds), 181–184. Pergamon, Oxford.
Anderson, F. and Hitchins, P.M. (1971). A radiotracking study of the black rhinoceros. *Journal of the South African Wildlife Management Association* **1**, 26–35.
Amlaner, C.J. (1978). Biotelemetry from free ranging animals. In "Animal Marking: Recognition Marking of Animals in Research" (B. Stonehouse, ed), 205–228. Macmillan, London.
Amlaner, C.J. (1980). The design of antennas for use in radio telemetry. In "A Handbook on Biotelemetry and Radio Tracking" (C.J. Amlaner and D.W. Macdonald, eds), 251–261. Pergamon, Oxford.
Amlaner, C.J. and Macdonald, D.W. (1980). A practical guide to radio tracking. In "A Handbook on Biotelemetry and Radio Tracking" (C.J. Amlaner and D.W. Macdonald, eds), 143–159. Pergamon, Oxford.
Amlaner, C.J., Sibly, R. and McCleery, R. (1978). Effects of transmitter weight on breeding success in herring gulls. *Biotelemetry and Patient Monitoring* **5**, 154–163.
Amstrup, S.C. (1980). A radio collar for game birds. *Journal of Wildlife Management* **44**, 214–217.
ARRL (1984). "The ARRL Antenna Book," 14th edition. American Radio Relay League, Newington, Connecticut.
Bailey, G.N.A., Linn, I.J. and Walker, P.J. (1973). Radioactive marking of small mammals. *Mammal Review* **3**, 11–23.
Banks, E.M., Brooks, R.J. and Schnell, J. (1975). A radiotracking study of home range and activity of the brown lemming (*Lemmus trimucronatus*). *Journal of Mammalogy* **56**, 888–901.
Barbour, R.W., Harvey, M.J. and Hardin, J.W. (1969). Home range, movements, and activity of the eastern worm snake, *Carphophis amoenus*. *Ecology* **50**, 470–476.
Bart, J. and Robson, D.S. (1982). Estimating survivorship when the subjects are visited periodically. *Ecology* **63**, 1078–1090.
Beale, D.M. and Smith, A.D. (1973). Mortality of pronghorn antelope fawns in western Utah. *Journal of Wildlife Management* **37**, 343–352.
Beaty, D.W. and Swapp, M.C. (1978). "Antenna Considerations for Biomedical Telemetry." Wildlife Society, Montana.
Bertram, B. (1980). The Serengeti radio-tracking program, 1971–1973. In "A Handbook on Biotelemetry and Radio Tracking" (C.J. Amlaner and D.W. Macdonald, eds), 625–631. Pergamon, Oxford.
Buechner, H.K., Craighead, F.C., Craighead, J.J. and Cote, C.E. (1971). Satellites for research on free roaming animals. *BioScience* **2**, 1201–1205.

Birks, J.D.S. and Linn, I.J. (1982). Studies of home range of the feral mink, *Mustela vison*. In "Telemetric Studies of Vertebrates" (C.L. Cheeseman and R.B. Mitson, eds), 231–257. Academic Press, London.

Boag, D.A. (1972). Effect of radio packages on behaviour of captive red grouse. *Journal of Wildlife Management* **36**, 511–518.

Bohus, B. (1974). Telemetered heart rate response of the rat during free and learned behaviour. *Biotelemetry* **1**, 193–201.

Boyd, J.C. and Sladen, W.J.L. (1971). Telemetry studies of the internal body temperatures of Adelie and emperor penguins at Cape Crozier, Ross Island, Antarctica. *Auk* **88**, 366–380.

Brand, C.J., Vowles, R.H. and Keith, L.B. (1975). Snowshoe hare mortality monitored by telemetry. *Journal of Wildlife Management* **39**, 741–747.

Brander, R.B. (1968). A radio-package harness for game birds. *Journal of Wildlife Management* **32**, 630–632.

Brander, R.B. and Cochran, W.W. (1971). Radio-location telemetry. In "Wildlife Management Techniques" (R.H. Giles, ed.), 95–103. The Wildlife Society, Washington.

Bray, O.E. and Corner, G.W. (1972). A tail clip for attaching transmitters to birds. *Journal of Wildlife Management* **36**, 640–642.

Broekhuizen, S., Van't Hoff, C.A., Jansen, M.B. and Niewold, F.J.J. (1980). Application of radio tracking in wildlife research in the Netherlands. In "A Handbook on Biotelemetry and Radio Tracking" (C.J. Amlaner and D.W. Macdonald, eds), 65–84. Pergamon, Oxford.

Brown, W.S. and Parker, W.S. (1976). Movement ecology of *Coluber constrictor* near communal hibernacula. *Copeia* (1976), 225–242.

Bub, H. and Oelke, H. (1980). "Markierungsmethoden für Vögel." Die Neue Brehm-Bücherei, Wittenberg-Lutherstadt.

Buchler, E.R. (1976). Chemiluminescent tag for tracking bats and other small nocturnal animals. *Journal of Mammalogy* **57**, 173–176.

Butler, R.W. and Jennings, J.G. (1980). Radio tracking of dolphins in the eastern tropical Pacific using VHF and HF equipment. In "A Handbook on Biotelemetry and Radio Tracking" (C.J. Amlaner and D.W. Macdonald, eds), 757–759. Pergamon, Oxford.

Calhoun, J.B. and Casby, J.U. (1958). "Calculation of Home Range and Density of Small Mammals." United States Public Health Service, Public Health Monograph 55.

Cederlund, G. and Lemnell, P.A. (1980). A simplified technique for mobile radio tracking. In "A Handbook on Biotelemetry and Radio Tracking" (C.J. Amlaner and D.W. Macdonald, eds), 319–322. Pergamon, Oxford.

Cederlund, G., Dreyfert, T. and Lemnell, P.A. (1979). "Radio-Tracking Techniques and the Reliability of Systems Used for Larger Birds and Mammals." Swedish Environmental Protection Board, Solna.

Charles-Dominique, P. (1977). Urine marking and territoriality in *Gallago alleni* (Waterhouse 1837 — Lorisidae Primates) — a field study by radio telemetry. *Zeitschrift für Tierpsychologie* **43**, 113–138.

Cheeseman, C.L. and Mallinson, P.J. (1980). Radio tracking in the study of bovine tuberculosis in badgers. In "A Handbook on Biotelemetry and Radio Tracking" (C.J. Amlaner and D.W. Macdonald, eds), 649–656. Pergamon, Oxford.

Cheeseman, C.L. and Mitson, R.B. (eds) (1982). "Telemetric Studies of Vertebrates." Symposia of the Zoological Society of London 49. Academic Press, London.

Church, K.E. (1980). Expanded radio tracking potential in wildlife investigations with the use of solar transmitters. In "A Handbook on Biotelemetry and Radio Tracking" (C.J. Amlaner and D.W. Macdonald, eds), 247–250. Pergamon, Oxford.

Chute, F.S., Fuller, W.A., Harding, P.J.R. and Herman, T.B. (1974). Radio-tracking of small mammals using a grid of overhead wires. *Canadian Journal of Zoology* **52**, 1481–1488.

Clarke, P.J. and Evans, F.C. (1954). Distance to nearest neighbour as a measure of spatial relationships in populations. *Ecology* **35**, 445–453.

Clute, R.K. and Ozoga, J.J. (1983). Icing of transmitter collars on white-tailed deer fawns. *Wildlife Society Bulletin* **11**, 70–71.

Coah, R.S., White, M., Trainer, D.O. and Glazener, W.C. (1971). Mortality of young white-tailed deer fawns in south Texas. *Journal of Wildlife Management* **35**, 47–56.

Cochran, W.W. (1967). "154–160 MHz Beacon (Tag) Transmitter for Small Animals." American Institute of Biological Sciences Bioinstrumentation Advisory Council Information Module 15.

Cochran, W.W. (1980). Wildlife telemetry. In "Wildlife Management Techniques Manual," 4th edition (S.D. Schemnitz, ed.), 507–520. Wildlife Society, Washington.

Cochran, W.W. and Lord, R.D. (1963). A radio-tracking system for wild animals. *Journal of Wildlife Management* **27**, 9–24.

Cochran, W.W., Warner, D.W., Tester, J.R. and Keuchle, V.B. (1965). Automatic radio-tracking system for monitoring animal movements. *BioScience* **15**, 98–100.

Cooper, H.M. and Charles-Dominique, P. (1985). A microcomputer data acquisition-telemetry system: a study of activity in the bat. *Journal of Wildlife Management* **49**, 850–854.

Cox, D.R. (1972). Regression models and life tables. *Journal of the Royal Statistical Society* B **34**, 187–220.

Dalke, P.D. and Sime, P.R. (1938). Home and seasonal ranges of the eastern cottontail in Connecticut. *Transcripts of the North American Wildlife Conference* **3**, 659–669.

Davis, J.R., Von Relum, A.F., Smith, D.D. and Guynn, D.C. (1984). Implantable telemetry in beaver. *Wildlife Society Bulletin* **12**, 322–324.

Deat, A., Mauget, C., Mauget, R., Maurel, D. and Sempere, A. (1980). The automatic, continuous and fixed radio tracking system of the Chizé Forest: theoretical and practical analysis. In "A Handbook on Biotelemetry and Radio Tracking" (C.J. Amlaner and D.W. Macdonald, eds), 439–451. Pergamon, Oxford.

Dice, L.R. and Clark, P.J. (1953). "The Statistical Concept of Home Range as Applied to the Recapture of the Deermouse (*Peromyscus*)." University of Michigan Laboratory of Vertebrate Biology Contributions 62.

Dixon, K.R. and Chapman, J.A. (1980). Harmonic mean measure of animal activity areas. *Ecology* **61**, 1040–1044

Don, B.A.C. and Reynolls, K. (1983). A home range model incorporating biological attraction points. *Journal of Animal Ecology* **52**, 69–81.

Downhower, J.F. and Pauley, J.D. (1970). Automatic recordings of body temperature from free-ranging yellow-bellied marmots. *Journal of Wildlife Management* **34**, 639–641.

Dunn, J.E. (1979). A complete test for dynamic territorial interaction. *Proceedings of the Second International Conference on Wildlife Biotelemetry*, 159–169.

Dunn, J.E. and Gipson, P.S. (1977). Analysis of radio telemetry data in studies of home range. *Biometrics* **33**, 85–101.
Dunstan, T.C. (1972). A harness for radio-tagging raptorial birds. *Inland Bird Banding News* **44**, 4–8.
Dunstan, T.C. (1973). A tail feather package for radio-tagging raptorial birds. *Inland Bird Banding News* **45**, 3–6.
Eagle, T.C., Choromanski-Norris, J. and Keuchle, V.B. (1984). Implanting radio transmitters in mink and Franklin's ground squirrels. *Wildlife Society Bulletin* **12**, 180–184.
Eliassen, E. (1960). A method for measuring the heart rate and stroke/pulse pressures of birds in normal flight. *Årbok Universitet Bergen, Matematisk Naturvitenskapelig* **12**, 1–22.
Fitch, H.S. and Shirer, H.W. (1971). A radiotelemetric study of spatial relationships in some common snakes. *Copeia* (1971), 118–128.
Fitzner, R.E. and Fitzner, J.N. (1977). A hot melt glue technique for attaching radiotransmitter tail packages to raptorial birds. *North American Bird Bander* **2**, 56–57.
Folk, G.E. and Folk, M.A. (1980). Physiology of large mammals by implanted radio capsules. In "A Handbook on Biotelemetry and Radio Tracking" (C.J. Amlaner and D.W. Macdonald, eds), 33–43. Pergamon, Oxford.
Ford, G. and Krumme, D.W. (1979). The analysis of space use patterns. *Journal of Theoretical Biology* **76**, 125–155.
Fullagar, P.J. (1967). The use of radio-telemetry in Australian biological research. *Proceedings of the Ecological Society of Australia* **2**, 16–26.
Fuller, M.R. (1975). A technique for holding and handling raptors. *Journal of Wildlife Management* **39**, 824–825.
Fuller, M.R. and Tester, J.R. (1973). An automated radio tracking system for biotelemetry. *Raptor Research* **7**, 105–106.
Fuller, M.R., Levanon, N., Strikwerda, T.E., Seegar, W.S., Wall, J., Black, H.D., Ward, F.P., Howey, P.W. and Partelow, J. (1984). Feasibility of a bird-borne transmitter for tracking via satellite. *Biotelemetry* **8**, 375–378.
Garshelis, D.L. and Siniff, D.B. (1983). Evaluation of radio-transmitter attachments for sea otters. *Wildlife Society Bulletin* **11**, 378–383.
Gautier, J.P. (1980). Biotelemetry of the vocalizations of a group of monkeys. In "A Handbook on Biotelemetry and Radio Tracking" (C.J. Amlaner and D.W. Macdonald, eds), 535–544. Pergamon, Oxford.
Gessamen, J.A. (1974). Telemetry of electrocardiograms from free-living birds: a method of electrode placement. *Condor* **76**, 479–481.
Gillingham, M.P. and Bunnell, F.L. (1985). Reliability of motion-sensitive radio collars for estimating activity of black-tailed deer. *Journal of Wildlife Management* **49**, 951–958.
Gilmer, D.S., Keuchle, V.B. and Ball, I.J. (1971). A device for monitoring radio-marked animals. *Journal of Wildlife Management* **35**, 829–832.
Gilmer, D.S., Bell, J.J., Cowardin, L.M. and Reichmann, J.H. (1974). Effects of radio packages on wild ducks. *Journal of Wildlife Management* **38**, 243–252.
Gilmer, D.S., Cowardin, L.M., Duval, R.L., Mechlin, L.M., Shaiffer, C.W. and Keuchle, V.B. (1981). "Procedures for the Use of Aircraft in Wildlife Biotelemetry Studies." United States Department of the Interior, Fish and Wildlife Service Resource Publication 140.
Godfrey, G.A. (1970). A transmitter harness for small birds. *Inland Bird Banding News* **2**, 3–5.

Goldberg, J.S. and Haas, W. (1978). Interactions between mule deer dams and their radio-collared and unmarked fawns. *Journal of Wildlife Management* **42**, 422–425.

Göransson, G. (1980). Animal activity recorded by radio tracking and an audio time lapse recorder. In "A Handbook on Biotelemetry and Radio Tracking" (C.J. Amlaner and D.W. Macdonald, eds), 457–460. Pergamon, Oxford.

Graber, R.R. and Wunderle, S.L. (1966). Telemetric observations of a robin (*Turdus migratorius*). *Auk* **83**, 674–677.

Greager, D.C., Jenness, C.A. and Ward, G.D. (1979). An acoustically sensitive transmitter for telemetering the activities of wild animals. *Journal of Wildlife Management* **43**, 1001–1007.

Greenwood, R.J. and Sargeant, A.B. (1973). Influence of radio packs on captive mallards and blue-winged teal. *Journal of Wildlife Management* **37**, 3–9.

Hallberg, D.L., Janza, F.J. and Trapp, G.R. (1974). A vehicle-mounted directional antenna system for biotelemetry monitoring. *California Fish and Game* **60**, 172–177.

Harden Jones, F.R. and Arnold, G.P. (1982). Acoustic telemetry and the marine fisheries. In "Telemetric Studies of Vertebrates" (C.L. Cheeseman and R.B. Mitson, eds), 75–93. Academic Press, London.

Hardy, A.R. and Taylor, K.D. (1980). Radio tracking of *Rattus norvegicus* on farms. In "A Handbook on Biotelemetry and Radio Tracking" (C.J. Amlaner and D.W. Macdonald, eds), 657–665. Pergamon, Oxford.

Harris, S. (1980). Home ranges and patterns of distribution of foxes (*Vulpes vulpes*) in an urban area, as revealed by radio tracking. In "A Handbook on Biotelemetry and Radio Tracking" (C.J. Amlaner and D.W. Macdonald, eds), 685–690. Pergamon, Oxford.

Harrison, J.L. (1958). Range and movements of some Malayan rats. *Journal of Mammalogy* **39**, 190–206.

Harvey, M.J. and Barbour, R.W. (1965). Home range of *Microtus ochrogaster* as determined by a minimum area method. *Journal of Mammalogy* **46**, 398–402.

Haynes, J.M. (1978). Movement and Habitat Studies of Chinook Salmon and White Sturgeon. Ph.D. Thesis, University of Minnesota.

Hayes, R.W. (1982). A telemetry device to monitor big game traps. *Journal of Wildlife Management* **46**, 551–553.

Hayne, D.W. (1949). Calculation of size of home range. *Journal of Mammalogy* **30**, 1–18.

Heezen, K.L. and Tester, J.R. (1967). Evaluation of radio-tracking by triangulation with special reference to deer movements. *Journal of Wildlife Management* **31**, 124–141.

Heisey, D.M. and Fuller, T.K. (1985). Evaluation of survival and cause-specific mortality rates using telemetry data. *Journal of Wildlife Management* **49**, 668–674.

Hines, J.E. and Zwickel, F.C. (1985). Influence of radio packages on young blue grouse. *Journal of Wildlife Management* **49**, 1050–1054.

Hirons, G.J.M. and Owen, R.B. (1982). Radio tagging as an aid to the study of woodcock. In "Telemetric Studies of Vertebrates" (C.L. Cheeseman and R.B. Mitson, eds), 139–152. Academic Press, London.

Horton, G.I. and Causey, M.K. (1984). Brood abandonment by radio-tagged american woodcock hens. *Journal of Wildlife Management* **48**, 606–607.

Howey, P.W., Witlock, D.R., Fuller, M.R., Seegar, W.S. and Ward, F.P. (1984). A computerised biotelemetry receiving and datalogging system. *Biotelemetry* **8**, 442–446.

Huempfner, R.A., Maxson, S.J., Erickson, J. and Schuster, R.J. (1975). Recaptur-

ing radio-tagged ruffed grouse by nightlighting and snow burrow netting. *Journal of Wildlife Management* **39**, 821–823.
Ireland, L.C. (1980). Homing behaviour of juvenile green turtles *Chelonia mydas*. In "A Handbook on Biotelemetry and Radio Tracking" (C.J. Amlaner and D.W. Macdonald, eds), 761–764. Pergamon, Oxford.
Ireland, L.C. and Kanwisher, J.W. (1978). Underwater acoustic biotelemetry: procedures for obtaining information on the behaviour and physiology of free-swimming aquatic animals in their natural environments. In "The Behaviour of Fish and Other Aquatic Animals" (D.I. Mostofsky, ed.), 341–379. Academic Press, New York.
Jackson, D.H., Jackson, L.S. and Seitz, W.K. (1985). An expandable drop-off transmitter harness for young bobcats. *Journal of Wildlife Management* **49**, 46–49.
Jenkins, D. (1980). Ecology of otters in northern Scotland. I. Otter (*Lutra lutra*) breeding and dispersion in mid-Deeside, Aberdeenshire in 1974–79. *Journal of Animal Ecology* **49**, 713–735.
Jennrich, R.J. and Turner, F.B. (1969). Measurement of non-circular home range. *Journal of Theoretical Biology* **22**, 227–237.
Johnsen, P.B. (1980). The use of longterm ultrasonic implants for the location and harvest of schooling fish. In "A Handbook on Biotelemetry and Radio Tracking" (C.J. Amlaner and D.W. Macdonald, eds), 777–780. Pergamon, Oxford.
Johnson, R.N. and Berner, A.H. (1980). Effects of radio transmitters on released cock pheasants. *Journal of Wildlife Management* **44**, 686–689.
Kalbfleisch, J.D. and Prentice, R.L. (1980). "The Statistical Analysis of Failure Time Data." John Wiley & Sons, New York.
Keith, L.B., Meslow, E.C. and Rongstad, O.J. (1968). Techniques for snowshoe hare population studies. *Journal of Wildlife Management* **32**, 801–812.
Kenward, R.E. (1976). The effect of predation by goshawks, *Accipiter gentilis*, on woodpigeon, *Columba palumbus*, populations. D.Phil. Thesis, University of Oxford.
Kenward, R.E. (1977). Predation on released pheasants (*Fasianus colchicus*) by goshawks (*Accipiter gentilis*) in central Sweden. *Viltrevy* **10**, 79–112.
Kenward, R.E. (1978). Radio transmitters tail-mounted on hawks. *Ornis Scandinavica* **9**, 220–223.
Kenward, R.E. (1980). Radio monitoring birds of prey. In "A Handbook on Biotelemetry and Radio Tracking" (C.J. Amlaner and D.W. Macdonald, eds), 97–104. Pergamon, Oxford.
Kenward, R.E. (1982a) Techniques for monitoring the behaviour of grey squirrels by radio. In "Telemetric Studies of Vertebrates" (C.L. Cheeseman and R.B. Mitson, eds), 175–196. Academic Press, London.
Kenward, R.E. (1982b). Goshawk hunting behaviour, and range size as a function of food and habitat availability. *Journal of Animal Ecology* **51**, 69–80.
Kenward, R.E. (1985). Ranging behaviour and population dynamics in grey squirrels. In "Behavioural Ecology. Ecological Consequences of Adaptive Behaviour" (R.M. Sibly and R.H. Smith, eds), 319–330. Blackwell Scientific Publications, Oxford.
Kenward, R.E. (1987). Radio tagging and the analysis of animal movements. In "Microcomputers in Environmental Biology" (J.N.R. Jeffers, ed.), Chapter 6. IRL Press, Oxford.
Kenward, R.E., Marcström, M. and Karlbom, M. (1981). Goshawk winter ecology in Swedish pheasant habitats. *Journal of Wildlife Management* **45**, 397–408.
Kenward, R.E., Hirons, G.J.M. and Ziesemer, F. (1982). Devices for telemetering

the behaviour of free-living birds. In "Telemetric Studies of Vertebrates" (C.L. Cheeseman and R.B. Mitson, eds), 129–136. Academic Press, London.

Keuchle, V.B. (1982). State of the art of biotelemetry in North America. In "Telemetric Studies of Vertebrates" (C.L. Cheeseman and R.B. Mitson, eds), 1–18. Academic Press, London.

Ko, W.H. (1980). Power sources for implant telemetry and stimulation systems. In "A Handbook on Biotelemetry and Radio Tracking" (C.J. Amlaner and D.W. Macdonald, eds), 225–245. Pergamon, Oxford.

Koeppl, J.W., Slade, N.A. and Hoffmann, R.S. (1975). A bivariate home range model with possible application to ethological data analysis. *Journal of Mammalogy* **56**, 81–90.

Koeppl, J.W., Slade, N.A., Harris, K.S. and Hoffmann, R.S. (1977). A three-dimensional home range model. *Journal of Mammalogy* **58**, 213–220.

Kolz, A.L. (1975). Mortality sensing wildlife transmitter. *Instrument Society of America Biomedical Sciences Institution Symposium* **12**, 57–60.

Kolz, A.L. (1983). Radio frequency assignments for wildlife telemetry: a review of the regulations. *Wildlife Society Bulletin* **11**, 56–59.

Kolz, A.L. and Corner, G.W. (1975). A 160-Megahertz telemetry transmitter for birds and bats. *Western Bird Bander* **50**, 38–40.

Kolz, A.L. and Johnson, R.E. (1975). An elevating mechanism for mobile receiving antennas. *Journal of Wildlife Management* **39**, 819–820.

Kolz, A.L. and Johnson, R.E. (1981). The human hearing response to pulsed-audio tones: implications for wildlife telemetry design. *Proceedings of the Third International Conference on Wildlife Biotelemetry*, 27–34.

Kolz, A.L., Lentfer, J.W. and Fallek, H.G. (1980). In "A Handbook on Biotelemetry and Radio Tracking" (C.J. Amlaner and D.W. Macdonald, eds), 743–752. Pergamon, Oxford.

Korschgen, C.E., Maxson, S.J. and Keuchle, V.B. (1984). Evaluation of implanted radio transmitter in ducks. *Journal of Wildlife Management* **46**, 982–987.

Kruuk, H. (1978). Spatial organisation and territorial behaviour of the European badger *Meles meles*. *Journal of Zoology, London* **184**, 1–19.

Kruuk, H., Gorman, M. and Parish, T. (1979). The use of ^{65}Zn for estimating populations of carnivores. *Oikos* **34**, 206–208.

Kruuk, H., Parish, T., Brown, C.A.J. and Carrera, J. (1979). The use of pasture by the European badger (*Meles meles*). *Journal of Applied Ecology* **16**, 453–459.

Laird, L.M. and Oswald, R.L. (1975). A note on the use of benzocaine (Ethyl p-amino benzoate) as a fish anaesthetic. *Fishery Management* **6**, 92–94.

Lance, A.N. and Watson, A. (1977). Further tests of radio-marking on red grouse. *Journal of Wildlife Management* **41**, 579–592.

Lawson, K., Kanwisher, J. and Williams, T.C. (1976). A UHF radio-telemetry system for wild animals. *Journal of Wildlife Management* **40**, 360–362.

Lee, J.E., White, G.C., Garrott, R.A., Bartmann, R.M. and Alldredge, A.W. (1985). Accessing accuracy of a radiotelemetry system for estimating animal locations. *Journal of Wildlife Management* **49**, 658–663.

Lemnell, P.A. (in press). Recent development of automatic radio tracking and transmission of coded information in wildlife research. *Proceedings of the XVIIth Congress of the International Union of Game Biologists*.

Lemnell, P.A., Johnsson, G., Helmersson, H., Holmstrand, O. and Norling, L. (1983). An automatic radio-telemetry system for position determination and data acquisition. *Proceedings of the Fourth International Conference on Wildlife Biotelemetry*, 76–93.

Le Munyan, C.D., White, W., Nybert, E. and Christian, J.J. (1959). Design of a miniature radio transmitter for use in animal studies. *Journal of Wildlife Management* **23**, 107–110.
Leuze, C.C.K. (1980). The application of radio tracking and its effect on the behavioural ecology of the water vole, *Arvicola terrestris* (Lacépède). In "A Handbook on Biotelemetry and Radio Tracking" (C.J. Amlaner and D.W. Macdonald, eds), 361–366. Pergamon Press, Oxford.
Linn, I.J. (1978). Radioactive techniques for small mammal marking. In "Animal Marking: Recognition Marking of Animals in Research" (B. Stonehouse, ed.), 177–191. Macmillan, London.
Linn, I.J. and Wilcox, P. (1982). A semi-automated system for collecting data on the movements of radio tagged voles. In "Telemetric Studies of Vertebrates" (C.L. Cheeseman and R.B. Mitson, eds), 197–205. Academic Press, London.
Lotimer, J.S. (1980). A versatile coded wildlife transmitter. In "A Handbook on Biotelemetry and Radio Tracking" (C.J. Amlaner and D.W. Macdonald, eds), 185–191. Pergamon Press, Oxford.
Loughlin, T.R. (1980). Radio telemetric determination of the 24-hour feeding activities of sea otters *Enhydra lutris*. In "A Handbook on Biotelemetry and Radio Tracking" (C.J. Amlaner and D.W. Macdonald, eds), 717–724. Pergamon Press, Oxford.
Lovett, J.W. and Hill, E.P. (1977). A transmitter syringe for recovery of immobilised deer. *Journal of Wildlife Management* **41**, 313–315.
Macdonald, D.W. (1978). Radio-tracking: some applications and limitations. In "Animal Marking: Recognition Marking of Animals in Research" (B. Stonehouse, ed), 192–204. Macmillan, London.
Macdonald, D.W. and Amlaner, C.J. (1980). A practical guide to radio tracking. In "A Handbook on Biotelemetry and Radio Tracking" (C.J. Amlaner and D.W. Macdonald, eds), 143–159. Pergamon Press, Oxford.
Macdonald, D.W., Ball, F.G. and Hough, N.G. (1980). The evaluation of home range size and configuration using radio tracking data. In "A Handbook on Biotelemetry and Radio Tracking" (C.J. Amlaner and D.W. Macdonald, eds), 405–424. Pergamon Press, Oxford.
McGinnis, S.M. (1967). The adaptation of biotelemetry to small reptiles. *Copeia* (1967), 472–473.
Mackay, R.S. (1964). Galapagos tortoise and marine iguana deep core body temperature measured by radio-telemetry. *Nature* **204**, 4956.
Madsen, T. (1984). Movements, home range size and habitat use of radio-tracked grass snakes (*Natrix natrix*) in southern Sweden. *Copeia* (1984), 707–713.
Marquès, M. (1972). Ensemble de reception 72 MHz à deux ariens permettant la localisation d'émetteurs destinés au radiotracking animaux. *Mammalia* **36**, 299–304.
Marquiss, M. and Newton, I. (1982). Habitat preference in male and female sparrowhawks (*Accipiter nisus*). *Ibis* **124**, 324–328.
Marshall, W.H. and Kupa, J.J. (1963). Development of radio-telemetry techniques for ruffed grouse studies. *Transactions of the North American Wildlife and Natural Resources Conference* **28**, 443–456.
Martin, M.L. and Bider, J.R. (1978). A transmitter attachment for blackbirds. *Journal of Wildlife Management* **42**, 683–685.
Mate, B.R., Harvey, J.T., Hobbs, L. and Maiefski, R. (1983). A new attachment device for radio-tagging large whales. *Journal of Wildlife Management* **47**, 868–872.

Mazurkiewicz, M. (1971). Shape, size and distribution of home ranges of *Clethrionomys glareolus* (Schreber, 1780). *Acta Theriologica* **16**, 23–60.
Mech, L.D. (1967). Telemetry as a technique in the study of predation. *Journal of Wildlife Management* **31**, 492–496.
Mech, L.D. (1974). Current techniques in the study of elusive wilderness carnivores. *International Congress on Game Biology* **11**, 315–322.
Mech, L.D. (1980). Making the most of radio-tracking: a summary of wold studies in Northeastern Minnesota. In "A Handbook on Biotelemetry and Radio Tracking" (C.J. Amlaner and D.W. Macdonald, eds), 85–95. Pergamon Press, Oxford.
Mech, L.D. (1983). "Handbook of Animal Radio-Tracking." University of Minnesota.
Mech, D.L., Chapman, R.C., Cochran, W.W., Simmons, L. and Seal, U.S. (1984). Radio-triggered anaesthetic dart collar for recapturing large mammals. *Wildlife Society Bulletin* **12**, 69–74.
Melquist, W.E. and Hornocker, M.G. (1979). Development and use of a telemetry technique for studying river otter. *Proceedings of the Second International Conference on Wildlife Biotelemetry*, 104–114.
Melvin, S.M., Drewien, R.C., Temple, S.A. and Bizeau, E.G. (1983). Leg-band attachment of transmitters for large birds. *Wildlife Society Bulletin* **11**, 282–285.
Michener, M.C. and Walcott, C. (1966). Navigation of single homing pigeons: airplane observations by radio tracking. *Science* **154**, 410–413.
Miles, M.A., De Souza, A.A. and Povoa, M.M. (1981). Mammal tracking an nest location in Brazilian forest with an improved spool-and-line device. *Journal of Zoology, London* **195**, 331–347.
Mitson, R.B., Storeton-West, T.J. and Pearson, N.D. (1982). Trials of an acoustic transponding fish tag compass. *Biotelemetry Patient Monitoring* **9**, 69–79.
Mohr, C.O. (1947). Table of equivalent populations of North American small mammals. *American Midland Naturalist* **37**, 223–249.
Mohr, C.O. and Stumpf, W.A. (1966). Comparison of methods for calculating areas of animal activity. *Journal of Wildlife Management* **30**, 293–304.
Mohus, I. (1983). Temperature telemetry from small birds. *Ornis Scandinavica* **14**, 273–277.
Morris, P. (1980). An elementary guide to practical aspects of radio tracking mammals. In "A Handbook on Biotelemetry and Radio Tracking" (C.J. Amlaner and D.W. Macdonald, eds), 161–168. Pergamon Press, Oxford.
Neft, D.S. (1966). "Statistical Analysis for Areal Distributions." Regional Scientific Research Institute Monograph Series 2.
Nenno, E.S. and Healey, W.M. (1979). Effects of radio packages on behaviour of wild turkey hens. *Journal of Wildlife Management* **43**, 760–765.
Nicholls, T.H. and Warner, D.W. (1968). A harness for attaching radio transmitters to large owls. *Bird Banding* **39**, 209–214.
Nolan, J.W., Russell, R.H. and Anderka, F. (1984). Transmitters for monitoring Aldrich snares set for grizzly bears. *Journal of Wildlife Management* **48**, 942–945.
Odum, E.P. and Kuenzler, E.J. (1955). Measurement of territory and home range size in birds. *Auk* **72**, 128–137.
Ogden, J. (1985) The California condor — capture and radio telemetry. *International Council for Bird Preservation Technical Publication* **5**, 475–476.
Osgood, D.W. (1970). Thermoregulation in water snakes studied by telemetry. *Copeia* (1970), 568–571.

Osgood, D.W. (1980). Temperature sensitive telemetry applied to studies of small mammal activity patterns. In "A Handbook on Biotelemetry and Radio Tracking" (C.J. Amlaner and D.W. Macdonald, eds), 525–528. Pergamon Press, Oxford.

Pages, E. (1975). Etude éco-éthologique du Pangolin *Manis tricuspis* à l'aide de la technique du radio-tracking. *Mammalia* **39**, 613–641.

Parish, T. (1980). A collapsible dipole antenna for radio tracking on 102 MHz. In "A Handbook on Biotelemetry and Radio Tracking" (C.J. Amlaner and D.W. Macdonald, eds), 263–268. Pergamon Press, Oxford.

Parish, T. and Kruuk, H. (1982). The uses of radio tracking combined with other techniques in studies of badger ecology in Scotland. In "Telemetric Studies of Vertebrates" (C.L. Cheeseman and R.B. Mitson, eds), 291–299. Academic Press, London.

Patric, E.F. and Serenbetz, R.W. (1971). A new approach to wildlife position finding telemetry. *New York Fish and Game Journal* **18**, 1–14.

Patton, D.R., Beaty, D.W. and Smith, R.H. (1973). Solar panels: an energy source for radio transmitters on wildlife. *Journal of Wildlife Management* **37**, 236–238.

Pearson, N.D. (1986). Automated telemetry systems. In "Animal Telemetry in the Next Decade", 49–50. Summaries of papers from a meeting organized by the Fisheries Laboratory, Lowestoft. Ministry of Agriculture, Fisheries and Food.

Perry, M.C. (1981). Abnormal behaviour of canvasbacks equipped with radio transmitters. *Journal of Wildlife Management* **45**, 786–789.

Perry, M.C., Haas, G.H. and Carpenter, J.W. (1981). Radio transmitters for mourning doves: a comparison of attachment techniques. *Journal of Wildlife Management* **45**, 524–527.

Philo, L.M., Follman, E.H. and Reynolds, H.V. (1981). Field surgical techniques for implanting temperature-sensitive transmitters in grizzly bears. *Journal of Wildlife Management* **45**, 772–775.

Priede, I.G. (1980). An analysis of objectives in telemetry studies of fish in the natural environment. In "A Handbook on Biotelemetry and Radio Tracking" (C.J. Amlaner and D.W. Macdonald, eds), 105–118. Pergamon Press, Oxford.

Priede, I.G. (1986). Satellite systems and tracking of animals. In "Animal Telemetry in the Next Decade", 51–56. Summaries of papers from a meeting organized by the Fisheries Laboratory, Lowestoft. Ministry of Agriculture, Fisheries and Food.

Proud, J.C. (1969). Wild turkey studies in New York by radio-telemetry. *New York Fish and Game Journal* **16**, 46–83.

Raim, A. (1978). A radio transmitter attachment for small passerine birds. *Bird Banding* **49**, 326–332.

Ramakka, J.M. (1972). Effects of radio-tagging on breeding behaviour of male woodcock. *Journal of Wildlife Management* **36**, 1309–1312.

Randolf, S.E. (1977). Changing spatial relationships in a population of *Apodmus sylvaticus* with onset of breeding. *Journal of Animal Ecology* **46**, 653–676.

Rasmussen, D.R. and Rasmussen, K.L. (1979). Social ecology of adult males in a confined troop of Japanese macaques (*Macaca fuscata*). *Animal Behaviour* **27**, 434–445.

Rawson, K.S. and Hartline, P.H. (1964). Telemetry of homing behaviour by the deermouse, *Peromyscus*. *Science* **146**, 1596–1598.

Reeve, N.J. (1980). A simple and cheap radio tracking system for use on hedgehogs. In "A Handbook on Biotelemetry and Radio Tracking" (C.J. Amlaner and D.W. Macdonald, eds), 169–173. Pergamon Press, Oxford.

Ricci, J.-C. and Vogel, P. (1984). Nouvelle méthode d'étude en nature des relations spatiales et sociales chez *Crocidura russula* (Mammalia, Soricidae). *Mammalia* **48**, 281–286.

Robinson, B.J. (1986). Data storage tags. In "Animal Telemetry in the Next Decade", 57–59. Summaries of papers from a meeting organized by the Fisheries Laboratory, Lowestoft. Ministry of Agriculture, Fisheries and Food.

Roth, H.U. (1983). Diet activity of a remnant population of European brown bears. *International Conference on Bear Research and Management* **5**, 223–229.

Salz, D. and Alkon, P.U. (1985). A simple computer-aided method for estimating radio-location error. *Journal of Wildlife Management* **49**, 664–668.

Sanderson, G.C. and Sanderson, B.C. (1964). Radio-tracking rats in Malaya — a preliminary study. *Journal of Wildlife Management* **28**, 752–768.

Sawby, S.W. and Gessamen, J.A. (1974). Telemetry of electrocardiograms from free-living birds: a method of electrode placement. *Condor* **76**, 479–481.

Sayre, M.W., Baskett, T.S. and Blenden, P.B. (1981). Effects of radio tagging on breeding behaviour of mourning doves. *Journal of Wildlife Management* **45**, 428–434.

Schubauer, J.P. (1981). A reliable radio-telemetry tracking system suitable for studies of chelonians. *Journal of Herpetology* **15**, 117–120.

Schwartz, A., Weaver, J.D., Scott, N.R. and Cade, T.J. (1977). Measuring the temperature of eggs during incubation under captive falcons. *Journal of Wildlife Management* **41**, 12–17.

Schweinsberg, R.E. and Lee, L.J. (1982). Movement of four satellite-monitored polar bears in Lancaster Sound, Northwest Territories. *Arctic* **35**, 504–511.

Serveen, C., Thier, T.T., Jonkel, C.J. and Beatty D. (1981). An ear-mounted transmitter for bears. *Wildlife Society Bulletin* **9**, 56–57.

Seidensticker, J.C., Hornocker, M.G., Knight, R.R. and Judd, S.L. (1970). "Equipment and Techniques for Radiotracking Mountain Lions and Elk." University of Idaho Forest and Range Experimental Station Bulletin 6.

Sibly, R.M. and McCleery, R.H. (1980). Continuous observation of individual herring gulls during the incubation season using radio tags: an evaluation of the technique and a cost-benefit analysis of transmitter power. In "A Handbook on Biotelemetry and Radio Tracking" (C.J. Amlaner and D.W. Macdonald, eds), 345–352. Pergamon Press, Oxford.

Siegfried, W.R., Frost, P.G.H., Ball, I.J. and McKinney, D.F. (1977). Effects of radio packages on African black ducks. *South African Journal of Wildlife Research* **7**, 37–40.

Siegel, S. (1957). "Non-Parametric Statistics." McGraw-Hill, New York.

Sievert, P.R. and Keith, L.B. (1985). Survival of snowshoe hares at a geographic range boundary. *Journal of Wildlife Management* **49**, 854–866.

Siniff, D.B. and Tester, J.R. (1965). Computer analysis of animal movement data obtained by telemetry. *BioScience* **15**, 104–108.

Siniff, D.B., Tester, J.R. and Keuchle, V.B. (1969). Population studies of Weddell seals at McMurdo Station. *Antarctic Journal of the United States* **4**, 120–121.

Skiffins, R.M. (1982). Regulatory control of telemetric devices used in animal studies. In "Telemetric Studies of Vertebrates" (C.L. Cheeseman and R.B. Mitson, eds), 19–30. Academic Press, London.

Skirnisson, K. and Feddersen, D. (1985). Erfahrungen mit der Implantation von Sendern bei freilebenden Steinmardern. *Zeitschrift für Jagdwissenschaft* **30**, 228–235.

Small, R.J. and Rusch, D.H. (1985). Backpacks versus ponchos: survival and

movements of radio-marked ruffed grouse. *Wildlife Society Bulletin* **13**, 163–165.
Smith, E.N. (1974). Multichannel temperature and heart rate radio-telemetry transmitter. *Journal of Applied Physiology* **36**, 252–255.
Smith, E.N. (1980). Physiological radio telemetry of vertebrates. In "A Handbook on Biotelemetry and Radio Tracking" (C.J. Amlaner and D.W. Macdonald, eds), 45–55. Pergamon Press, Oxford.
Smith, E.N. and Worth, D.J. (1980). Atropine effect on fear bradycardia of the eastern cottontail rabbit. In "A Handbook on Biotelemetry and Radio Tracking" (C.J. Amlaner and D.W. Macdonald, eds), 549–555. Pergamon Press, Oxford.
Smith, H.R. (1980). Growth, reproduction and survival in *Peromyscus leucopus* carrying intraperitoneally implanted transmitters. In "A Handbook on Biotelemetry and Radio Tracking" (C.J. Amlaner and D.W. Macdonald, eds), 367–374. Pergamon Press, Oxford.
Smith, R.M. and Trevor-Deutsch, B. (1980). A practical, remotely-controlled, portable radio telemetry receiving apparatus. In "A Handbook on Biotelemetry and Radio Tracking" (C.J. Amlaner and D.W. Macdonald, eds), 269–273. Pergamon Press, Oxford.
Smits, A.W. (1984). Activity patterns and thermal biology of the toad *Bufoboreas halophilus*. *Copeia* (1984), 689–696.
Snedecor, G.W. and Cochran, W.G. (1971). "Statistical Methods." 2nd edition. Iowa State University Press, Ames.
Snyder, W.D. (1985). Survival of radio-marked hen ring-necked pheasants in Colorado. *Journal of Wildife Management* **49**, 1044–1050.
Solomon, D.J. and Storeton-West, T.J. (1983). "Radio Tracking of Migratory Salmonids in Rivers: Development of an Effective System." Ministry of Agriculture, Fisheries and Food, Fisheries Research Technical Report 75.
Southwood, T.R.E. (1966). "Ecological Methods." Methuen, London.
Spencer, W.D. and Barrett, R.H. (1984). An evaluation of the harmonic mean measure for defining carnivore activity areas. *Acta Zoologica Fennica* **171**, 255–259.
Springer, J.T. (1979). Some sources of bias and sampling error in radio triangulation. *Journal of Wildife Management* **43**, 926–935.
Standora, E.A. (1977). An eight-channel radio telemetry system to monitor alligator body temperature in a heated reservoir. In "Proceedings of the First International Conference on Wildlife Biotelemetry" (F.M. Long, ed.), 70–78. University of Wyoming.
Standora, E.A., Spotila, J.R., Keinath, J.A. and Shoop, C.R. (1984). Body temperatures, diving cycles, and movement of a subadult leatherback turtle, *Dermochelys coriacea*. *Herpetologica* **40**, 169–176.
Stasko, A.B. and Pincock, D.G. (1977). Review of underwater biotelemetry, with emphasis on ultrasonic techniques. *Journal of the Fisheries Research Board of Canada* **34**, 1262–1285.
Stebbings, R.E. (1982). Radio tracking greater horseshoe bats with preliminary observations on flight patterns. In "Telemetric Studies of Vertebrates" (C.L. Cheeseman and R.B. Mitson, eds), 161–173. Academic Press, London.
Stebbings, R.E. (1986). Bats. In "Animal Telemetry in the Next Decade", 20–22. Summaries of papers from a meeting organized by the Fisheries Laboratory, Lowestoft. Ministry of Agriculture, Fisheries and Food.
Stickel, L.F. (1954). A comparison of certain methods of measuring ranges of small mammals. *Journal of Mammalogy* **35**, 1–15.
Stoddart, L.C. (1970). A telemetric method of detecting jackrabbit mortality. *Journal of Wildlife Management* **34**, 501–507.

Stonehouse, B. (ed.) (1978). "Animal Marking: Recognition Marking of Animals in Research." Macmillan, London.

Stouffer, R.H., Gates, J.E., Holutt, L.H. and Stauffer, J.R. (1983). Surgical implantation of a transmitter package for radio-tracking endangered hellbenders. *Wildlife Society Bulletin* **11**, 384–386.

Strathearn, S.M., Lotimer, J.S., Kolenosky, G.B. and Lintack, W.M. (1984). An expanding break-away collar for black bear. *Journal of Wildlife Management* **48**, 939–942.

Swanson, G.A. and Keuchle, V.B. (1976). A telemetry technique for monitoring waterfowl activity. *Journal of Wildlife Management* **40**, 187–189.

Swihart, R.K. and Slade, N.A. (1985). Testing for independence of observations in animal movements. *Ecology* **66**, 1176–1184.

Swingland, I.R. and Frazier, J.G. (1980). The conflict between feeding and overheating in the Aldabran giant tortoise. In "A Handbook on Biotelemetry and Radio Tracking" (C.J. Amlaner and D.W. Macdonald, eds), 269–273. Pergamon Press, Oxford.

Taylor, K.D. and Lloyd, H.G. 1978. The design, construction and use of a radio-tracking system for some British mammals. *Mammal Review* **8**, 117–141.

Tester, J.R. (1971). Interpretation of ecological and behavioural data on wild animals obtained by telemetry with special reference to errors and uncertainties. *Proceedings of a Symposium on Biotelemetry*, 385–408. CSER, Pretoria, South Africa.

Thomas, D.W. (1980). Plans for a lightweight inexpensive radio transmitter. In "A Handbook on Biotelemetry and Radio Tracking" (C.J. Amlaner and D.W. Macdonald, eds), 175–179. Pergamon Press, Oxford.

Timko, R.E. and Kolz, A.L. (1982). Satellite sea turtle tracking. *Marine Fisheries Review* **44**, 19–24.

Tonkin, J.M. (1983). Ecology of the Red Squirrel (*Sciurus vulgaris*) in Mixed Woodland. Ph.D. Thesis, University of Bradford.

Trent, T.T. and Rongstad, O.J. (1974). Home range and survival of cottontail rabbits in southwestern Wisconsin. *Journal of Wildlife Management* **38**, 459–472.

Uda, S. and Mushiake, Y. (1954). "Yagi-Uda Antenna." Sasaki Printing and Publishing, Sendai, Japan (cited in Beaty and Swapp, 1978).

Verts, B.J. (1963). Equipment and techniques for radiotracking striped skunks. *Journal of Wildlife Management* **27**, 325–339.

Voight, D.R. and Broadfoot, J. (1983). Locating pup-rearing dens of red foxes with radio-equipped woodchucks. *Journal of Wildlife Management* **47**, 858–859.

Voight, D.R. and Tinline, R.R. (1980). Strategies for analyzing radio tracking data. In "A Handbook on Biotelemetry and Radio Tracking" (C.J. Amlaner and D.W. Macdonald, eds), 387–404. Pergamon Press, Oxford.

Warner, R.E. and Etter, S.L. (1983). Reproduction and survival of radio-marked hen ring-necked pheasants. *Journal of Wildlife Management* **47**, 369–375.

Watkins, W.A., Moore, K.E., Wartzok, D. and Johnson, J.H. (1981). Radio-tracking of finback (*Balaenoptera physalus*) and humpback (*Megaptera novaeangliae*) whales in Prince William Sound, Alaska. *Deep-Sea Research* **28**A, 577–588.

Weeks, R.W., Long, F.M., Lindsay, J.E., Bailey, R., Patula, D. and Green M. (1977). Fish tracking from the air. *Proceedings of the First International Conference on Wildlife Telemetry*, 63–69.

Whitehouse, S. (1980). Radio tracking in Australia. In "A Handbook on Biotelemetry and Radio Tracking" (C.J. Amlaner and D.W. Macdonald, eds) 733–739. Pergamon Press, Oxford.

Whitehouse, S.J.O. and Steven, D. (1977). A technique for aerial radio tracking. *Journal of Wildlife Management* **41**, 771–775.

Widén, P. (1982). Radio monitoring the activity of goshawks. In "Telemetric Studies of Vertebrates" (C.L. Cheeseman and R.B. Mitson, eds), 153–160. Academic Press, London.

Williams, T.C. and Williams, J.M. (1967). Radio tracking of homing bats. *Science* **155**, 1435–1436.

Williams, T.C. and Williams, J.M. (1970). Radiotracking of homing and feeding flights of a neotropical bat, *Phyllostomus hastatus*. *Animal Behaviour* **18**, 302–309.

Winter, J.D., Keuchle, V.B., Siniff, D.B. and Tester, J.R. (1978). "Equipment and Methods for Tracking Freshwater Fish." Agricultural Experimental Station of the University of Minnesota Miscellaneous Report 152.

Wolcott, T.G. (1980a). Optical and radio optical techniques for tracking nocturnal animals. In "A Handbook on Biotelemetry and Radio Tracking" (C.J. Amlaner and D.W. Macdonald, eds), 333–338. Pergamon Press, Oxford.

Wolcott, T.G. (1980b). Heart rate telemetry using micropower integrated circuits. In "A Handbook on Biotelemetry and Radio Tracking" (C.J. Amlaner and D.W. Macdonald, eds), 279–286. Pergamon Press, Oxford.

Wood, D.A. (1976). Squirrel collars. *Journal of Zoology, London* **180**, 513–518.

Wooley, J.B. and Owen, R.B. (1978). Energy costs of activity and daily energy expenditure in the black duck. *Journal of Wildlife Management* **42**, 739–745.

Yerbury, M.J. (1980). Long range tracking of *Crocodylus porosus* in Arnhem Land, Northern Australia. In "A Handbook on Biotelemetry and Radio Tracking" (C.J. Amlaner and D.W. Macdonald, eds), 765–776. Pergamon Press, Oxford.

Zicus, M.C., Schultz, D.F. and Cooper, J.A. (1983). Canada goose mortality from neckband icing. *Wildlife Society Bulletin* **11**, 70–71.

Ziesemer, F. (1981). Methods of assessing predation. In "Understanding the Goshawk" (R.E. Kenward and I.M. Lindsay, eds), 144–151. International Association of Falconry and Conservation of Birds of Prey, Oxford.

Zimmermann, F., Geraud, H. and Charles-Dominique, P. (1975). Le radiotracking des vertebrates: conseils et techniques d'utilisation. *La Terre et la Vie* **30**, 309–346.

Zinnel, K.A. and Tester, J.R. (1984). Non-intrusive monitoring of plains pocket gophers. *Bulletin of the Ecological Society of America* **65**, 166.

Appendix I

This appendix lists some suppliers whose equipment has been used in radio tagging projects, without implying any endorsement of their products.

Acoustic tags and 104 MHz radio equipment

D.W. McKay, 120 Dell Road, Lowestoft, Suffolk, United Kingdom.

Radio tags and receiving equipment

Advanced Telemetry Systems, Inc., 23859 NE Highway 65, Bethel, Minnesota 55005, U.S.A.

Austec Electronics Ltd, 1006, 11025 — 82 Avenue, Edmonton, Alberta, Canada T6G 0T1.

AVM Instrument Company, Ltd, 2368 Research Drive, Livermore, California 94550, U.S.A.

Biotrack, Stoborough Croft, Grange Road, Wareham, Dorset BH20 5AJ, United Kingdom.

Custom Electronics of Urbana, Inc., 2009 Silver Ct. W., Urbana, Illinois 61801, U.S.A.

Karl Wagener, Herwarthstrasse 22, 5000 Köln, West Germany.

Mariner Radar (Lowestoft) Ltd, Bridle Way, Camps Heath, Lowestoft, Suffolk, United Kingdom.

Telemetry Systems, Inc., P.O. Box 187, Mequon, Wisconsin 53092, U.S.A.

Televilt HB, PL 5226, 71050 Storå, Sweden.

Telonics, Inc., 1300 W. University, Mesa, Arizona 85201, U.S.A.

Tracktech AB, Lundagatan 5, S-171 63 Solna, Sweden.

Wildlife Materials, Inc., R.R.1, Carbondale, Illinois 62901, U.S.A.

Passive transponders

Fish Eagle Trading Company, Little Faringdon Mill, Lechlade, Gloucestershire, United Kingdom.

Recording equipment

Squirrel Data Loggers, Grant Instruments, Cambridge, Ltd, Barrington, Cambridge CB2 1BR, United Kingdom.

Rustrak Chart Recorders, Gulton Industries Inc., Industrial Service Laboratories, 4354 Olive Street, St. Louis, Montana 63108, U.S.A.

Repeater compasses

Marinex Marine Marketing Ltd., 11 Balena Close, Creekmore, Poole, Dorset BH17 7DB, United Kingdom.

Spectrum analysers

Fieldtech Heathrow Ltd., Huntavia House, 420 Bath Road, Longford, West Drayton UB7 0LL, United Kingdom.

Telescopic masts

Clark Masts Ltd., Binstead, Isle of Wight PO33 3PA, United Kingdom.

Appendix II

Addresses of firms supplying components and materials which have proved satisfactory in radio tags. The addresses are all in the United Kingdom, unless otherwise indicated.

Beta lights

Saunders-Roe Developments Ltd, Millington Road, Hayes, Middlesex UB3 4NB.

Crystals

IQD Ltd, North Street, Crewkerne, Somerset TA18 7JU.

Hy-Q International (UK) Ltd, Station Road, Whittlesford, Cambridge CB2 4NL.

General electronic components, tools and mercury cells

Farnell Electronic Components, Ltd, Canal Road, Leeds, Yorkshire LS12 2TU.

RS Components Ltd, PO Box 99, Corby, Northants NN17 9RS.

Lithium cells

Saft (UK) Ltd, Castle Works, Station Road, Hampton, Middlesex TW12 2BI.

Tadiran, Israel Electronics Industries Ltd, 3 Hashalom Road, P.O. Box 648, Tel-Aviv, Israel.

Tecnicad Ltd (Sanyo agent), 20/22 Poole Hill, Bournemouth, Dorset BH20 5PS.

Mercury and silver cells

Rayovac Ltd, Galleon House, King Street, Maidstone, Kent ME14 1BG.

Mercury switches

Gentech International Ltd, Grangestone Industrial Estate, Girvan, Ayrshire KA26 9PS.

D.P. Distribution Ltd, Unit A, Sadds Industrial Estate, Clacton on Sea, Essex CO16 6BZ.

Microminiature resistors

Lemo (UK) Ltd, 12 North Street, Worthing, Sussex BN11 1DU.

Steatite-Roederstein Ltd, Hagley House, Hagley Road, Birmingham B16 8QW.

Polystyrene tubes

Arnold R. Horwell Ltd, 2 Grangeway, Kilburn Highway, London NW6 2BP.

Sarstedt (UK) Ltd, 165 Scudamore Road, Leicester, Leics. LE3 1UQ.

Redcap miniature plate ceramic capacitors

STC Mercator, South Denes, Great Yarmouth, Norfolk NR30 3PX.

Specialist heat-shrinkable tubing

Rayfast Ltd, Unit 24, Ganton Way, Techno Trading Estate, Swindon, Wilts. SN2 6EZ.

Sutures (absorbable)

Ethicon Ltd, PO Box 408, Bankhead Avenue, Edinburgh EH11 4HE, Scotland.

Tantalum capacitors

Corning Ltd, PO Box 37, Pallion Trading Estate, Sunderland, Tyne & Wear SR4 6SU.

Toroids & cores

Neosid Ltd, Brownfields, Welwyn Garden City, Herts AL7 1AN.

Transistors

Celdis Ltd, 37–39 Loverock Road, Reading, Berks RG3 1ED.

STC Multicomponent, Edinburgh Way, Harlow, Essex CM20 2DE.

Trimmers, tantalum capacitors

Tekelec Components Ltd, Cumberland House, Baxter Avenue, Southend on Sea, Essex SS2 6FA.

Zinc–air cells

Gould Micro-Power Products, 11 Ash Road, Wrexham Industrial Estate, Wrexham, Clwyd LL13 9UF.

Appendix III

The three range analysis programs in this appendix can be used for mononuclear convex polygons which follow several "coring" rules (Program 1), for multinuclear convex polygons derived by cluster analysis (Program 2), and for harmonic mean contouring (Program 3). Further programs of this type can be found in Kenward (1987).

All these programs are intended mainly as demonstrations of the techniques, and to provide BASIC algorithms which may be converted for use on machines other than BBC Microcomputers. They are not intended for serious data analysis, although they can be used for this if you are content to process small numbers of ranges at a time: the use of data statements limits the amount of material that can be held in memory with each program.

If you have large quantities of data to analyse, or an aversion to typing and debugging long program listings, you may prefer to buy a copy of the RANGES II disc from Biotrack (see Appendix I). RANGES II is a suite of menu-driven programs, including versions of those listed here but with other routines for entering data and preparing disc files, for analysing distances, speeds and headings between consecutive fixes, for overlap and habitat analyses, for plots of range size against number of fixes, and for plotting multirange utilization distributions.

A. Data and displays

The following three programs have several features in common. They use the same data format and plotting sequences, and a number of other algorithms, at the same line numbers. If you start by entering, debugging and saving Program 1, you can use it as a basis for Programs 2 and 3, by deleting a few lines from it and adding the necessary new ones.

The data are entered in sets from line 1000. The first variable, read at line 155, is a string variable which labels the range. It is best to keep strings short, or they may plot across the display on the right of the screen. Following the label, there is a pair of coordinates for a nest or den. If an animal has no nest, or this is not to be shown in the analyses, you must enter $-999,-999$ as a missing value code. After the nest coodinates or missing value code, fix coordinates are entered in order, east before north, and each range terminates with $-1,-1,-1,-1$. Thus:

1000DATAJF7,23,45,34,38,......,35,42,$-1,-1,-1,-1$

would be for juvenile female 7, nest at 23E,45N, first fix at 34E,38N, and last fix at 35E,42N. The next range must start at line 1050, and subsequent ranges at 50 line intervals. Coordinates should be a continous series from east to west and north to south. If your study area crosses a national grid boundary, such that it includes positions 999 and 001, you should enter the coordinates as 999 and 1001 (or 99 and 101 if your study area is small). A single -1 is used to terminate the whole set of range data.

The programs will run in MODE4 with 150 fixes in each range on the 32K BBC B. If you have more memory available, for instance on a BBC Master model, you can accommodate more fixes by changing the values in the arrays at line 90, and by changing MODE4 at line 95 to MODE132.

The variable "i" on line 100 is the resolution of your fixes. For instance, if your fixes are at 100 m intervals, you can save entering a lot of 0s by using e.g. 34,45 and 35,45 for coordinates which are 100 m apart, and setting $i=100$. If you enter coordinates to the nearest metre, then $i=1$. The areas and distances on the display will then be given correctly in hectares and metres.

The programs all make provision for printing out results, but if you want hardcopy of the graphics output you will need a screen dump routine or ROM. Insert the call for this routine or ROM at line 375. For instance, if you have the Computer Concepts PRINTMASTER ROM, line 375 should read:

375VDU5:MOVE0,2000:*GDUMP

Then, when you see the prompt "PRESS RETURN" at the end of each range, press 1 before <RETURN> to initiate the dump.

Program 1

Since these programs use almost all the memory available in MODE4, they have been compacted by using multi-line statements, with single letters for almost all variable and procedure labels. This makes them hard to follow. For those who wish to understand what they do, or to make changes, the Program 1 procedures are as follows:

PROCD 390–430 Controls the sequence of the analysis options, and progressively excludes fixes during core range analysis.

PROCE 435–495 Finds the polygon corner fixes. Starting at the most westerly (or northwesterly) fix, it seeks the fix which gives the maximum tangent in the first quadrant, until the most northerly (or northeasterly) fix is reached, at which time the flag R% is set to 1. The maximum fix which gives the maximum tangent in the second quadrant is then sought, until the most easterly (or southeasterly) fix is passed, and W% set to 1. This continues through the most southerly fix, setting S% to 1, until the most westerly point is reached again. The corner fixes are stored in X%(B%),Y%(B%), and the polygon edges, including the boundary strip, in X(C%),Y(C%).

PROCF 500–535 Finds the most westerly (F%,G%), easterly (I%,H%), northerly (L%,M%) and southerly (O%,P%) fixes. If there are several separate fixes with, say, the most westerly coordinates, then the most northern of these is chosen.

PROCT 540–595 Finds the maximum tangent in the relevent quadrant.

PROCt 600–610 Stores the fix with the maximum tangent as a corner fix.

PROCA 615–645 Divides each polygon into triangles, along diagonals from the first corner, and adds the areas of these triangles, which are found from the length of the sides by Simpson's rule.

PROCK 650–660 Draws the polygon sides to the corners of the boundary strip.

PROCZ 665–725 Lists the utilization distribution areas at the left of the screen, and plots them as a function of the proportion of fixes at the bottom right of the screen.

PROCX 730–750 Prints the coordinates of the activity centre, and of the polygon corner fixes.

PROCP 760–800 Finds the westmost (I%), eastmost (J%), southmost (K%) and northmost (L%) coordinates, and scales the plotting of fixes and polygon lines accordingly.

PROCN 885–895 Stores the mean (s), median (u) and maximum (v) distances of fixes from the nest.

PROCC 900–915 Calculates the coordinates of the arithmetic mean, and stores the mean, median and maximum distances from it.

PROCR 920–940 Ranks fixes by distance from the chosen activity centre.

PROCI 945–970 Finds the fix with the minimum inverse reciprocal mean distance to the other fixes, and stores mean, median and maximum distances from it.

PROCY 975–990 Stores the distance of each fix from the activity centre in D(A%). Plots, at the chosen activity centre, a circle for the nest, a large square for the activity centre, a small square for the focusing activity centre, or a diamond for the harmonic mean.

When you run Program 1, you will first be asked whether you wish to print results. You will then be asked to:

CHOOSE POLYGON PLOTS BASED ON:

0 Outer fixes only
1 The nest
2 The harmonic mean fix
3 The arithmetic mean position
4 The focusing arithmetic mean

If you select "0", the corner coordinates and area will be given for a minimum convex polygon plotted round all the fixes. If you select polygons based on the nest or other centre, you will then be asked:

CHOOSE:
0 To print edge coordinates
1 To plot utilization distribution

A third question asks you to select a percentage of points round which the polygon outlines should plot. If your choice for the first question was option 4 (focusing arithmetic mean), and you then choose a utilization distribution with plots at 5% intervals, you will see the small square (depicting this centre) move as the centre is recalculated after excluding each outermost fix. Typical output for this program is shown in Fig. 8.4 and 8.5.

Program 2

This program can be made from Program 1 by first deleting lines 10, 20–30, 55, 185, 205–225, 390–430, 675–685, 740–745 and 885–995. Then add the new lines 10, 55, 185, 205–210, 675–685, 740–745 and 810–995 from Program 3.
The new procedures in this program are:

PROCM 810–845 Finds the three fixes which are the smallest mean distance apart, and saves the distance I between the third fix and the closest two.

PROCQ 955–995 Flashes the chosen cluster off the screen, finds their nearest neighbour at distance L, and marks each fix with a small square to show that it is assigned.

PROCO 850–950 Checks whether the next cluster has a smaller I than the L for each existing cluster (stored in the array Q%). It then checks that the next fix(es) to be added are unassigned, merging clusters if they are. As each 5% of the fixes are added, it finds the area of probability polygons round all the clusters.

The questions at the start of this program are similar to those in Program 1, and the output is also similar, as shown in Fig. 8.7.

Program 3

To make this program from Program 1, delete all the lines *except* 5, 15, 45–55, 95–105, 140–170, 190–195, 370–385, 665–670, 685–725, 760–780, 790–800 and 950–960. Add from Program 3 the lines 10, 30–35, 90, 105–130, 175–185, 200–245, 390–645, 675–680, 785 and 805–975.
The new procedures are:

PROCY 940–975 Estimates the fix with the lowest inverse reciprocal mean distance to the others, and marks it with a diamond.

PROCM 390–420 Uses scaling data from PROCN to move each fix to the centre of the nearest grid square on a 40 × 40 grid, which extends five squares to the west and south of each range. It then estimates the inverse reciprocal mean distance to all the fixes from each intersection on the grid, multiplies the means by 1000 and stores them in the array X%(x,y), and records their minimum value as M%.

PROCI 570–580 Finds the corner values (E%,F%,G%,H%) and central value (W%) for each grid square from X%(x,y).

PROCG 615–645 Tests whether W% is within the current incrementing isoline value, M%, and counts any fixes in the square if it is, until the target number of fixes, D%, for the chosen percentage has been reached.

PROCJ 425–445 Moves east to west and south to north through the array X%(x,y), testing whether the corner values are all within or all outside the current isoline. If they are within the isoline, it increments the area by one square. If the isoline cuts the square it calls PROCQ.

PROCQ 450–535 Uses PROCF to interpolate the isoline on the edges and diagonals of the current square, and estimates the area proportion (E) of the square which lies within the line.

PROCF 540–550 See PROCQ, above.

PROCK 555–565 Finds the coordinates for the points where isolines cut edges and diagonals of each square.

PROCX 805–815 Finds the pixel coordinates for plotting isolines.

On running this program, you are given two analysis options. The first, "0", gives an enlarged plot of each range, which helps in distinguishing the individual isolines. The isolines plot at 10% intervals or at a single selected value. The option "1" gives a utilization distribution plot as in the previous programs.

The program first finds the fix with the minimum inverse reciprocal mean distance to the others, and marks this with a diamond. It then calculates the inverse reciprocal mean distance to all the fixes at each grid intersection, showing its progress at the bottom left of the screen. Finally, it increments the isoline value, plotting the line when the chosen percentage of fixes is contained, or at 10% intervals (5% for utilization distribution plots) if no fixed value was chosen. Typical output is shown in Fig. 8.8.

Program 1

```
50ONERROR:VDU3,4:CLS:REPORT:PRINT", line ";ERL"     ";:GOTO380
10CLS:PRINT'"MONONUCLEAR PROBABILITY POLYGONS"'"By R.E. Kenward, ITE Furzebrook"
15@%=&9:PRINT'''"ENTER 1 FOR PRINTER OUTPUT, ELSE RETURN":INPUTp%
20k%=0:c=1:CLS:PRINT'''"CHOOSE POLYGON PLOTS BASED ON:"'''"0 Outer fixes only"'"1 The nest"
25PRINT"2 The harmonic mean fix"'"3 The arithmetic mean position"'"4 The focusing arithmetic mean"
30INPUTd%:IFd%>4VDU7:GOTO20 ELSEIFd%=0GOTO85
35PRINT"CHOOSE:"'''"0 To display edges/edge coordinates"'"1 To plot utilization distribution"'
40INPUTk%:PRINT''"POLYGON PROBABILITIES:"''"Enter % for edges";:
45IFk%=1PRINT", or"ELSEc=2:PRINT" + coordinates, or"
50PRINT"<RETURN> for edges only, at ";c*5"% spacing":INPUTq%:z%=(100-q%)/5
85Z%=1000
90DIME%(150),N%(150),D(150),X(50),Y(50),XX(30),YX(30),F(50),G(50),V(20)
95@%=&9:MODE4:VDU28,0,31,39,31:VDU5
100:=10:REM FIX RESOLUTION IN METRES.
105VDU23,141,0,0,0,56,40,56,0,0,23,142,0,0,0,16,56,16,0,0
150RESTOREZ%:Z%=Z%+50:REM LINE NUMBER INTERVAL FOR DATA SETS.
155READA$:IFA$="-1"GOTO380
160READx%,y%:IE%(0)=x%:N%(0)=y%:A%=1:IFx%=-999A%=0
165REPEAT:READE%,N%
170E%(A%)=E%:N%(A%)=N%:IFE%<>-1A%=A%+1
175UNTILE%=-1:N%=A%:PROCP
180FORA%=0TON%-1:MOVE(E%(A%)-w%)*e+o,(N%(A%)-s%)*n+p:VDU142:NEXT
185IFd%<2ANDX%<>-999MOVE(x%-w%)*e+o,(y%-s%)*n+p:VDU111
190IFp%=1VDU2:PRINT''''
195MOVE0,990:PRINTA$" (N=";N%")"
200DX=0:@%=&407:VDU23,143,0,0,124,68,68,124,0,0,23,144,0,0,16,56,124,254,124,56,16
205IFd%=1ANDx%=-999VDU4,7:PRINT"NO NEST! ";:GOTO370
215IFd%=1A$="Nest"ELSEIFd%=2A$="Harmonic mean"ELSEIFd%=3A$=";A$="Arithmetic mean"ELSEA$="Focusing mean"
220PROCA:H2=E/I^2:@%=&20210:IFd%=0PROCD:MOVE0,950:PRINT"Area:";H2" ha."
225@%=&9:IFd%>0PROCD:IFk%=1PROCZ
370VDU3,4:PRINT"PRESS RETURN  ";:INPUT""A:IFA<>1GOTO95
375REM
380VDU4,7:PRINT"PRESS RETURN  ";:INPUT""A:MODE7:PRINT'''"PRESS f2 TO RUN AGAIN"'':*KEY2CLEAR:GOTO5:M
385END
390DEFPROCD
395@%=&407:DX=0:D%=0:h%=0:b=2:IFd%=1a=1:PROCN ELSEa=0:IFd%=2PROCI ELSEPROCC
400REPEAT:F=h%*N%/20:IFk%=0ANDq%>0F=N%*(100-q%)/100
405G=INT(F):IFF-G>.5G=G+1
410REPEAT:IFD%>0ANDd%=4PROCC
```

```
412D%=D%+1:UNTILD%>=G+1:D%=D%-1:PROCA:IFk%=0ORk%=1AND(q%=00Rh%=z%))PROCK
415V(h%)=E/i^2:IFk%=10Rq%=0:VDU3,4,12:PRINT"Polygon  %  ";(20-h%)*5;:VDU5:IFp%=1VDU2
420IFk%=0ANDq%>0MOVE0,950:PRINT";q%" area:"V(h%)" ha.":PROCX:h%=16
425h%=h%+c:UNTILh%=17:VDU4:PRINT:VDU5
430ENDPROC
435DEFPROCE
440PROCF:X%(0)=F%:Y%(0)=G%:X(0)=F%+.5:Y(0)=G%+.5:U%=F%:V%=G%:R%=0:W%=0:S%=0
445B%=1:C%=1:REPEAT:IFU%=L%ANDV%=M%R%=1:X(C%)=M%+.5:Y(C%)=M%+.5:C%=C%+1
450IFU%=H%ANDV%=I%ANDR%=1W%=1:X(C%)=H%+.5:Y(C%)=H%+.5:C%=C%+1:IFM%=P%X(C%)=0%-.5:Y(C%)=P%-.5:GOTO490
455IFU%=0%ANDV%=P%ANDW%=1S%=1:X(C%)=0%-.5:Y(C%)=0%-.5:C%=C%+1:IFO%=F%ANDP%=G%C%=C%-1:B%=B%-1
460IFO%=F%ANDP%=G%S%=0
465IFV%=M%ANDU%<L%X%(B%)=L%:Y%(B%)=M%:X(C%)=L%-.5:U%=L%:V%=M%:GOTO490
470IFV%>I%ANDU%=H%X%(B%)=H%:Y%(B%)=I%:Y(C%)=I%+.5:U%=H%:V%=I%:GOTO490
475IFV%=P%ANDU%>O%X%(B%)=O%:Y%(B%)=P%:X(C%)=O%+.5:Y(C%)=P%-.5:U%=O%:V%=P%:GOTO490
480IFV%<G%ANDU%=F%X%(B%)=F%:Y%(B%)=G%:Y(C%)=G%-.5:Y(C%)=F%-.5:U%=F%:V%=G%:GOTO490
485PROCT
490B%=B%+1:C%=C%+1:UNTILX(C%-1)=F%-.5ANDY(C%-1)=G%-.5:B%=B%-1:X(C%)=F%-.5:Y(C%)=G%+.5
495ENDPROC
500DEFPROCF
505M%=N%(0):G%=N%(0):I%=N%(0):P%=N%(0):L%=E%(0):F%=E%(0):H%=E%(0):O%=E%(0)
510FORA%=1TON%-D%-1:Q%=E%(A%):T%=N%(A%):IFQ%<F%OR(Q%=F%ANDT%>G%)F%=Q%:G%=T%
515IFQ%>H%OR(Q%=H%ANDT%<I%)H%=Q%:I%=T%
520IFT%>M%OR(T%=M%ANDQ%>L%)M%=T%:L%=Q%
525IFT%<P%OR(T%=P%ANDQ%<O%)P%=T%:O%=Q%
530NEXT
535ENDPROC
540DEFPROCT
545T=0:FORA%=0TON%-D%-1:Q%=E%(A%):T%=N%(A%):IFQ%=U%ANDT%=V%GOTO585
550IFR%=0ANDT%>V%t=(T%-V%)/(Q%-U%):IFt>T:g=-.5:h=-.5:PROCt:GOTO585
555IFR%=0GOTO585
560IFR%=1ANDW%=0ANDQ%>U%t=(Q%-U%)/(V%-T%):IFt>T:g=-.5:h=.5:PROCt:GOTO585
565IFR%=1ANDW%=0GOTO585
570IFW%=1ANDS%=0ANDT%<V%t=(V%-T%)/(U%-Q%):IFt>T:g=.5:h=.5:PROCt:GOTO585
575IFW%=1ANDS%=0GOTO585
580IFS%=1ANDQ%<U%t=(U%-Q%)/(T%-V%):IFt>T:g=.5:h=-.5:PROCt:GOTO585
585NEXT
590U%=E%(E%):V%=N%(E%)
595ENDPROC
600DEFPROCt
605E%=A%:X%(B%)=Q%:Y%(B%)=T%:X(C%)=Q%+g:Y(C%)=T%+h:T=t
610ENDPROC
```

Program 1 *Contd.*

```
615DEFPROCA
620PROCE:FORA%=1TOC%-1:F(A%)=SQR((Y(0)-Y(A%))^2+(X(0)-X(A%))^2)
625NEXT:FORA%=1TOC%-2:G(A%)=SQR((Y(A%)-Y(A%+1))^2+(X(A%)-X(A%+1))^2):NEXT:E=0
630FORA%=1TOC%-2:S=0.5*(G(A%)+F(A%)+F(A%+1)):IFS(G(A%)ORS(F(A%)ORS(F(A%+1)GOTO640
635E=E+SQR(S*(S-G(A%))*(S-F(A%))*(S-F(A%+1)))
640NEXT
645ENDPROC
650DEFPROCK
655MOVE(F%-.5-w%)*e+o+12,(G%+.5-s%)*n+p-15:FORA%=1TOC%:DRAW(X(A%)-w%)*e+o+12,(Y(A%)-s%)*n+p-15:NEXT
660ENDPROC
665DEFPROCZ
670S$="-based areas:":IFp%=1VDU2
675IFd%=1i%=78:S$="Nest"+S$ELSEIFd%=2i%=72:S$="Hc"+S$ELSEIFd%=3i%=84:S$="Ac"+S$ELSEi%=70:S$="Fc"
680V(0)=H2:E=H2/100:@%=&20209:MOVE0,930:PRINT'S$;" max =   ";H2" ha":VDU3:@%=&20110
685IFE=0GOTO710ELSEMOVE0,810:PRINT"%  ha  (%max)    ("V(A%)/E")   ":NEXT
690FORA%=1TO16:PRINTSTR$(100-A%*5);" "V(A%)" ("V(A%)/E")   ":NEXT
695MOVE1270,45:DRAW600,45:MOVE600,455:MOVE500,455:PRINT;" 100":MOVE500,255:PRINT;" 50"
700MOVE500,55:PRINT;" 0":FORA%=0TO4:MOVE(600+(A%*150)),25:PRINTSTR$(100-(20*A%)):NEXT
705MOVE520,355:PRINT"%A":FORA%=0TO16:MOVE600+(38*A%),55+(V(A%)*4/E):VDU(i%):NEXT
710IFE=0GOTO725ELSEIFp%=1VDU2
715MOVE0,2000:FORA%=0TO3:FORB%=1TO4:PRINTSTR$((20-B%-4*A%)*5)"%="; V(B%+4*A%)"ha.("V(B%+4*A%)/E") ";
720NEXTB%:PRINT:NEXTA%:VDU3
725ENDPROC
730DEFPROCX
735IFp%=1VDU2
740@%=&d07:MOVE0,320:IFd%>OPRINT'A$" at ";x","y
745@%=&9:PRINT';q%"% edges: ":FORA%=0TOB%-1:PRINT;X%(A%)","Y%(A%)"   ";:NEXT:PRINT':VDU3
750ENDPROC
760DEFPROCP
765I%=10^8:J%=0:K%=10^8:L%=0:FORA%=0TON%-1:IFE%(A%)<I%I%=E%(A%)
770IFE%(A%)>J%J%=E%(A%)
775IFN%(A%)<K%K%=N%(A%)
780IFN%(A%)>L%L%=N%(A%)
785NEXT:w%=I%:e%=J%:s%=K%:n%=L%:IFk%=0o=500:p=300ELSEo=700:p=500
790J%-J%:L%=L%-K%:IFJ%>L%S=J%/30 ELSES=L%/30
795e=(1100-o)/(S*30):n=(900-p)/(S*30)
800ENDPROC
885DEFPROCN
890a=1:x=x%:y=y%:PROCY:PROCR:s=L*i/(N%-1):u=D(INT(N%/2))*i:v=D(N%-1)*i
```

```
895ENDPROC
900DEFPROCC
905J%=0:K%=0:FORA%=0TON%-D%-1:J%=J%+E%(A%):K%=K%+N%(A%):NEXT:x=J%/(N%-D%):y=K%/(N%-D%)
910a=0:PROCY:PROCR:v=D(N%-D%-1)*i:u=D(INT((N%-D%)/2))*i:s=L*i/(N%-D%-1)
915ENDPROC
920DEFPROCR
925REPEAT:Q=0:FORA%=a  TON%-D%-b:W=D(A%):R=D(A%+1):Q=E%(A%):T%=N%(A%):U%=E%(A%+1):V%=N%(A%+1)
930D=W-R:IFD>0:E%(A%)=U%:N%(A%)=V%:E%(A%+1)=Q%:N%(A%+1)=T%:D(A%)=R:D(A%+1)=W:Q=Q+1
935NEXT:UNTILQ=0
940ENDPROC
945DEFPROCI
950A=0:FORA%=0TON%-1:B=0:FORB%=0TON%-1:IFA%=B%GOTO960
955IFE%(A%)=E%(B%)ANDN%(A%)=N%(B%)B=B+1 ELSEB=B+1/(SQR((E%(A%)-E%(B%))^2+(N%(A%)-N%(B%))^2))
960NEXTB%:IFB>A:Q=A%:A=B
965NEXTA%:x=E%(Q):y=N%(Q):PROCY:PROCR:v=D(N%-1)*i:u=D(INT(N%/2))*i:s=(N%-1)*i/A
970ENDPROC
975DEFPROCY
980L=0:FORA%=a  TON%-1:D(A%)=SQR((E%(A%)-x)^2+(N%(A%)-y)^2):L=L+D(A%):NEXT
985VDU5:MOVE(x-w%)*e+o,(y-s%)*n+p:IFd%=2VDU144ELSEIFd%=3VDU143ELSEIFd%=4VDU141
990ENDPROC
1000DATAAM9,-9/85,41,46,48,45,38,47,40,44,40,45,38,47,42,43,41,44,45,38
1010DATA48,46,49,47,48,49,48,44,44,44,51,40,49,45,51,39,47,48,49,40,45,50,49,42,51,44,51
1020DATA47,47,43,47,42,45,47,44,42,45,42,44,57,47,50,48,42,50,-1,-1
1050DATAAM7,-9/85,-999,48,45,41,44,36,44,47,48,49,48,50,29,49,51,48,30,52,41,43
1060DATA48,48,60,47,38,47,48,49,53,47,41,49,46,47,48,49,43,49,50,49,39,53
1070DATA41,51,50,48,40,50,40,47,40,47,41,46,36,51,40,47,40,47,35,51,-1,-1
1100DATAAF8-9/85,15,31,16,18,20,24,14,29,14,34,19,17,34,14,28,14,29
1110DATA19,38,15,27,18,34,17,34,15,31,17,31,20,41,15,29,15,31,28,34,21,27
1120DATA27,30,32,37,24,39,26,29,27,38,15,31,15,31,29,34,16,30,15,28,-1,-1
1150DATAAF2-9/85,35,34,43,32,34,32,34,32,52,48,35,53,36,31,33,26,41,37,39,29
1160DATA51,48,35,31,35,31,46,45,50,49,43,37,41,31,39,33,43,31,43,32,37,33
1170DATA37,33,61,47,34,33,36,33,62,47,39,31,34,32,35,31,35,32,-1,-1
1200DATAAF2-9/85,27,48,37,37,25,47,25,48,40,45,38,47,42,43,28,47,25,46,46,47
1210DATA26,46,28,44,27,47,28,43,26,45,20,40,29,43,30,39,26,46,41,50,24,46
1220DATA34,43,34,43,30,46,26,43,28,49,27,48,31,45,27,47,47,28,47,-1,-1
1250DATA-1
```

Program 2

```
50ONERROR:VDU3,4:CLS:REPORT:PRINT", line ";ERL" ";:GOTO380
10CLS:PRINT'""MULTINUCLEAR POLYGONS BY CLUSTERING"'"By R.E. Kenward, ITE Furzebrook"
15@%=&9:PRINT'""ENTER 1 FOR PRINTER OUTPUT, ELSE RETURN":INPUTp%
20c=1
35CLS:PRINT"CHOOSE:"'"0 To display edges/edge coordinates"'"1 To plot utilization distribution"'
40INPUTk%:PRINT'"POLYGON PROBABILITIES:"'"Enter % for edges";;
45IFk%=1PRINT", or"ELSEc=2:PRINT" + coordinates, or"
50PRINT"<RETURN> for edges only, at ";c*5"% spacing":INPUTq%:z%=(100-q%)/5
85Z%=1000:DIMDD(150),O%(150),P%(150),S%(75),Q%(75),R%(75),B%(2),K%(20)
90DIME%(150),N%(150),D(150),X(50),Y(50),X%(30),Y%(30),F(50),G(50),V(20)
95@%=&9:MODE4:VDU28,0,31,39,31:VDU5
100i=10:REM FIX RESOLUTION IN METRES.
105VDU23,141,0,0,0,56,40,56,0,0,23,142,0,0,0,16,56,16,0,0
150RESTOREZ%:Z%+=50:REM LINE NUMBER INTERVAL FOR DATA SETS.
155READA$:IFA$=""-1"GOTO380
160READx%,y%:E%(0)=x%:N%(0)=y%:O%(0)=x%:P%(0)=y%:A%=1:IFx%=-999A%=0
165REPEAT:READE%,N%
170E%(A%)=E%:N%(A%)=N%:O%(A%)=E%:P%(A%)=N%:IFE%<>-1A%=A%+1
175UNTILE%=-1:N%=A%:PROCP
180FORA%=0TON%-1:MOVE(E%(A%)-w%)*e+o,(N%(A%)-s%)*n+p:VDU142:NEXT
185IFx%<>-999MOVE(x%-w%)*e+o,(y%-s%)*n+p:VDU111
190IFp%=1VDU2:PRINT'''
195MOVE0,1020:PRINTA$"  (N=";N%")"
200D%=0:@%=&407:VDU3,23,143,0,16,40,68,130,68,40,16,23,144,0,126,66,66,66,126,0
205PROCA:H2=E/i^2:@%=&9:PROCQ:IFk%=1PROCz
370VDU3,4:PRINT"PRESS RETURN ";:INPUT'"A:IFA<>1GOTO95
375REM
380PRINT'""PRESS RETURN "; :INPUT'"A:MODE7:PRINT'""PRESS f2 TO RUN AGAIN"'':*KEY2CLEAR:GOTO5;M
385END
435DEFPROCE
440PROCF:X%(0)=F%:Y%(0)=G%:X(0)=F%:Y%(0)=G%+.5:U%=F%:V%=G%:R%=0:W%=0:S%=0
445B%=1:C%=1:REPEAT:IFU%=L%ANDV%=M%R%=1:X(C%)=L%+.5:Y(C%)=M%+.5:C%=C%+1
450IFU%=H%ANDV%=I%ANDR%=1W%=1:X(C%)=H%+.5:Y(C%)=I%-.5:C%=C%+1:IFM%=P%X(C%)=0%-.5:Y(C%)=P%-.5:GOTO490
455IFU%=0%ANDV%=P%ANDW%=1S%=1:X(C%)=0%-.5:Y(C%)=P%-.5:C%=C%+1:IF0%=F%ANDP%=G%X(C%)=C%-1:B%=B%-1
460IF0%=F%ANDP%=G%S%=0
465IFV%=M%ANDU%<L%X(C%)=L%+.5:Y(C%)=M%+.5:U%=L%:V%=M%:GOTO490
470IFV%>I%ANDU%=H%X(C%)=H%+.5:Y(C%)=I%-.5:U%=H%:V%=I%:GOTO490
475IFV%=P%ANDU%>0%X(C%)=0%-.5:Y(C%)=P%-.5:U%=0%:V%=P%:GOTO490
480IFV%<G%ANDU%=F%X(C%)=F%+.5:Y(C%)=G%-.5:U%=F%:V%=G%:GOTO490
```

```
485PROCT
490B%=B%+1:C%=C%+1:UNTILX(C%-1)=F%-.5ANDY(C%-1)=G%-.5:B%=B%-1:X(C%)=F%-.5:Y(C%)=G%+.5
495ENDPROC
500DEFPROCF
505M%=N%(O):G%=N%(O):I%=N%(O):P%=N%(O):L%=E%(O):F%=E%(O):H%=E%(O):O%=E%(O)
510FORA%=1TONZ-D%-1:Q%=E%(A%):T%=N%(A%):IFQ%<F%OR(Q%=F%ANDT%>G%)F%=Q%:G%=T%
515IFQ%>H%OR(Q%=H%ANDT%<I%)H%=Q%:I%=T%
520IFT%>M%OR(T%=M%ANDQ%>L%)M%=T%:L%=Q%
525IFT%<P%OR(T%=P%ANDQ%<O%)P%=T%:O%=Q%
530NEXT
535ENDPROC
540DEFPROCT
545T%=0:FORA%=0TON%-D%-1:Q%=E%(A%):T%=N%(A%):IFQ%=U%ANDT%=V%GOTO585
550IFR%=0ANDT%>V%t=(T%-V%)/(Q%-U%):IFt>T:g=-.5:h=.5:PROCt:GOTO585
555IFR%=0GOTO585
560IFR%=1ANDW%=0ANDQ%>U%t=(Q%-U%)/(V%-T%):IFt>T:g=.5:h=-.5:PROCt:GOTO585
565IFR%=1ANDW%=0GOTO585
570IFW%=1ANDS%=0ANDT%<V%t=(V%-T%)/(U%-Q%):IFt>T:g=.5:h=-.5:PROCt:GOTO585
575IFW%=1ANDS%=0GOTO585
580IFS%=1ANDQ%<U%t=(U%-Q%)/(T%-V%):IFt>T:g=-.5:h=-.5:PROCt:GOTO585
585NEXT
590U%=E%(E%):V%=N%(E%)
595ENDPROC
600DEFPROCt
605E%=A%:X%(B%)=Q%:Y%(B%)=T%:X(C%)=Q%+g:Y(C%)=T%+h:T=t
610ENDPROC
615DEFPROCA
620PROCE:FORA%=1TOC%-1:F(A%)=SQR((Y(O)-Y(A%))^2+(X(O)-X(A%))^2)
625NEXT:FORA%=1TOC%-2:G(A%)=SQR((Y(A%)-Y(A%+1))^2+(X(A%)-X(A%+1))^2):NEXT:E=0
630FORA%=1TOC%-2:S=0.5*(G(A%)+F(A%+1)):IFS<G(A%)ORS<F(A%+1)GOTO640
635E=E+SQR(S*(S-G(A%))*(S-F(A%))*(S-F(A%+1)))
640NEXT
645ENDPROC
650DEFPROCK
655MOVE(F%-.5-w%)*e+o+12,(G%+.5-s%)*n+p-15:FORA%=1TOC%:DRAW(X(A%)-w%)*e+o+12,(Y(A%)-s%)*n+p-15:NEXT
660ENDPROC
665DEFPROCZ
670IFp%=1VDU2
675S$="Clustered areas:":i%=77
680V(O)=H2:E=H2/100:@%=&20209:MOVEO,930:PRINT'S$'"max (mono) = ";H2" ha":VDU3:@%=&20110
685IFE=0GOTO710ELSEMOVEO,810:PRINT"% ha (%max) nuc."
```

Program 2 *Contd.*

```
690FORA%=1TO16:PRINTSTR$(100-A%*5);" "V(A%)" ("V(A%)/E") "STR$(K%(A%)):NEXT
695MOVE1270,45:DRAW600,45:DRAW600,455:MOVE500,255:PRINT;" 50"
700MOVE500,55:PRINT;"   0":FORA%=OTO4:MOVE(600+(A%*150),25:PRINTSTR$(100-(20*A%)):NEXT
705MOVE20,355:PRINT"%A":FORA%=OTO16:MOVE600+(38*A%),55+(V(A%)*4/E):VDU(i%):NEXT
710IFE=OGOTO725ELSEIFp%=1VDU2
715MOVE0,2000:FORA%=OTO3:FORB%=1TO4:PRINTSTR$((20-B%-4*A%)*5)"%=";V(B%+4*A%)"ha.("V(B%+4*A%)/E") ";
720NEXTB%:PRINT:NEXTA%:VDU3
725ENDPROC
730DEFPROCX
735IFp%=1VDU2
740@%=&9:MOVE0,370:IFd%=1PRINT';q%"% edges: "
745MOVE0,335-35*d%:FORA%=OTOB%-1:PRINT;X%(A%)",":Y%(A%)" ";:NEXT:PRINT:VDU3
750ENDPROC
760DEFPROCP
765I%=10^8:J%=0:K%=10^8:L%=0:FORA%=OTON%-1:IFE%(A%)<I%I%=E%(A%)
770IFE%(A%)>J%J%=E%(A%)
775IFN%(A%)<K%K%=N%(A%)
780IFN%(A%)>L%L%=N%(A%)
785NEXT:w%=I%:e%=J%:s%=K%:n%=L%:IFk%=0o=500:p=300ELSEo=700:p=500
790J%=J%-I%:L%=L%-K%:IFJ%>L%S=J%/30 ELSES=L%/30
795e=(1100-o)/(S*30):n=(900-p)/(S*30)
800ENDPROC
810DEFPROCM
815J%=10^10:FORI%=OTON%-1:IFDD(I%)>OGOTO840
820H=10^10:I=10^10:FORB%=OTON%-1
825D(B%)=SQR((0%(B%)-0%(I%))^2+(P%(B%)-P%(I%))^2):IFB%<>I%ANDD(B%)<H:H=D(B%):U%=B%
830NEXTB%:FORB%=OTON%-1:IFB%<>I%ANDB%<>U%ANDD(B%)<I:I=D(B%):V%=B%
835NEXTB%:IF(H+I)<J:J=H+I:B%(O)=I%:B%(1)=U%:B%(2)=V%:Q=I
840NEXTI%
845ENDPROC
850DEFPROC0
855VDU3:FORA%=OTON%-1:DD(A%)=O:NEXT:s=0:PROCM:s=s+1:c%=N%-3:h%=16
860FORA%=OTO2:DD(B%(A%))=s:NEXT:S%(1)=3:S%(O)=1:r%=s:PROCQ:REPEAT:F=h%*N%/20:IFk%=OANDq%>OF=N%*(100-q%)/100
865G=INT(F):IFF-G>.5:G=G+1
870REPEAT:IFc%<=G GOTO915ELSEPROCM:C%=0:FORA%=1TOs:IFQ>Q(A%):C%=1
875NEXT:IFC%>OGOTO895
880IFDD(B%(1))>ODD(B%(O))=DD(B%(1))=S%(DD(B%(1)))+1:r%=DD(B%(1)):PROCQ:c%=c%-1:GOTO915
885IFDD(B%(2))>ODD(B%(O))=DD(B%(2)):S%(DD(B%(2)))+1:r%=DD(B%(2)):PROCQ:c%=c%-1:GOTO915
890s=s+1:DD(B%(O))=s:DD(B%(1))=s:DD(B%(2))=s:S%(s)=3:S%(O)=s:r%=s:PROCQ:c%=c%-3:GOTO915
```

```
895C=10^10:FORA%=1TOs:IFQ(A%)<C:C=Q(A%):u%=A%
900NEXT:v%=DD(R%(u%)):IFv%=ODD(R%(u%))=u%:S%(u%)=S%(u%)+1:r%=u%:PROCQ:c%=c%-1:GOTO915
905FORA%=OTON%-1:IFDD(A%)=v%DD(A%)=u%
910NEXT:S%(u%)=S%(u%)+S%(v%):S%(v%)=0:Q(v%)=10^10:r%=u%:PROCQ
915UNTILc%<=G:IFk%=10Rq%=OVDU4,12:PRINT"Polygon % ";(20-h%)*5;:VDU5
920d%=0:AR=0:FORJ%=1TOS%(O):K%=0:FORI%=OTON%-1:IFDD(I%)=J%E%(K%)=O%(I%):N%(K%)=P%(I%):K%=K%+1
925NEXTI%:IFS%(J%)=OGOTO935ELSEd%=d%+1:D%=N%-K%:PROCA:AR=AR+E
930IFk%=OOR(k%=1AND(q%=OORh%=z%))PROCk:IFk%=OANDq%>OPROCX
935FORI%=OTON%-1:E%(I%)=O:N%(I%)=O:NEXTI%:NEXTJ%:K%(h%)=d%:V(h%)=AR/i^2
940IFk%=OANDq%>O@%=&407:VDU3-p%:MOVEO,950:PRINT;q%"% area:"V(h%)" ha.":VDU3:h%=0
945h%=h%-c:UNTILh%<O:VDU4:PRINT:VDU5
950ENDPROC
955DEFPROCQ
960GCOLO,0:FORA%=OTON%-1:IFDD(A%)=r%ANDK%<2MOVE(O%(A%)-w%)*e+o,(P%(A%)-s%)*n+p:VDU142
965NEXT:GCOL1,1:L=10^10:FORA%=OTON%-1:IFDD(A%)<>r%GOTO980
970FORB%=OTON%-1:IFDD(B%)<>r%D=SQR((O%(B%)-O%(A%))^2+(P%(B%)-P%(A%))^2):IFD<L:L=D:v%=B%
975NEXTB%
980NEXTA%:Q(r%)=L:R%(r%)=v%:v%=DD(v%)
985FORA%=OTON%-1:IFDD(A%)=r%ANDK%<2MOVE(O%(A%)-w%)*e+o,(P%(A%)-s%)*n+p:VDU141
990NEXT
995ENDPROC
1000DATAAM9-9/85,41,46,48,45,38,47,47,40,44,40,45,38,47,42,43,41,44,45,38
1010DATA48,46,49,49,48,45,44,50,49,45,51,39,47,48,49,40,45,50,49,42,51,44,51
1020DATA47,47,43,47,42,45,47,44,42,45,42,44,57,47,50,48,42,50,-1,-1
1050DATAAM7-9/85,-999,-999,48,45,41,44,36,47,48,49,48,50,29,49,51,48,30,52,41,43
1060DATA48,48,60,47,38,47,48,53,47,41,49,44,47,48,49,43,49,50,49,39,53
1070DATA41,51,50,48,40,50,40,47,40,47,41,46,36,51,40,47,40,47,35,51,-1,-1
1100DATAAF8-9/85,15,31,16,18,20,24,14,29,14,34,19,17,34,14,28,14,29
1110DATA19,38,15,27,18,34,17,34,15,31,17,31,20,41,15,29,15,31,28,34,21,27
1120DATA27,30,32,37,24,39,26,29,27,38,15,31,15,31,29,34,16,30,15,28,-1,-1
1150DATAAF2-9/85,35,34,43,32,34,32,34,32,52,48,35,53,36,31,33,26,41,37,39,29
1160DATA51,48,35,31,35,31,46,45,50,49,43,37,41,31,39,33,43,31,43,32,37,33
1170DATA37,33,61,47,34,33,36,33,62,47,39,31,34,32,35,31,35,32,-1,-1
1200DATAAF2-9/85,27,48,37,37,25,48,40,45,38,47,42,43,28,47,25,46,46,47
1210DATA26,46,28,44,27,47,28,43,26,45,20,40,29,43,30,39,26,46,41,50,24,46
1220DATA34,43,43,30,46,26,43,28,49,27,48,31,45,27,47,27,47,28,47,-1,-1
1250DATA-1
```

Program 3

```
50 ONERROR:VDU3,4:CLS:REPORT:PRINT", line ";ERL  ";:GOTO380
10 CLS:PRINT'"HARMONIC MEAN CONTOURING"'"By R.E. Kenward, ITE Furzebrook"
15 @%=&9:PRINT'"ENTER 1 FOR PRINTER OUTPUT, ELSE RETURN":INPUTp%
20 c=1
35 CLS:PRINT','"CHOOSE:",'"0 For enlarged single ranges"'"1 To plot utilization distribution"
40 INPUTk%:PRINT"CONTOUR PROBABILITIES",'"Enter % for a single isopleth, or":IFk%=0c=2
50 PRINT"<RETURN> for plots at ";c*5"% intervals":INPUTq%:z%=(100-q%)/5
85 Z%=1000
90 DIME%(150),N%(150),D%(150),X%(40,40),V(20)
95 @%=&9:MODE4:VDU28,0,31,39,31:VDU5
100 i=10:REM FIX RESOLUTION IN METRES.
105 VDU23,142,0,0,0,16,56,16,0,0,23,143,0,16,40,68,130,68,40,16
150 RESTOREZ%:Z%=Z%+50:REM LINE NUMBER INTERVAL FOR DATA SETS.
155 READA$:IFA$=""-1"GOTO380
160 READx%,y%:E%(O)=x%:N%(O)=y%:A%=1:IFx%=-999A%=0
165 REPEAT:READE%,N%
170 E%(A%)=E%:N%(A%)=N%:IFE%<>-1A%=A%+1
175 UNTILE%=-1:N%=A%:PROCP
180 FORA%=0TON%-1:MOVE(E%(A%)-w%)*e+o,(N%(A%)-s%)*n+p:VDU142:NEXT
185 IFx%<>-999MOVE(x%-w%)*e+o,(y%-s%)*n+p:VDU111
190 IFp%=1VDU2:PRINT'''
195 MOVE0,1020:PRINTA$"    (N=";N%")"'
200 VDU3:@%=&408:PROCy:PROCM:PROCR
205 D%=0:h%=16:REPEAT:F=(20-h%)*N%/20:IFk%=0ANDq%>0F=N%*q%/100
210 G=INT(F):IFF-G>.5G=G+1
215 REPEAT:D%=D%+1:UNTILD%>=G:IFk%=0ANDq%>0G=q%ELSEG=(20-h%)*5
225 M%=D%(D%-1)+1:VDU3,4:PRINT';M%*i/1000"m Isoline, "G"% Isopleth";:VDU5
230 T=0:FORA%=0TO39:FORB%=0TO39:PROCJ:NEXTB%:NEXTA%:V(h%)=T*S^2/i^2:IFp%=1VDU2
235 IFk%=0ANDq%>0MOVE0,950:PRINT';q%"% area:"V(h%)" ha.":h%=0
240 h%=h%-c:UNTILh%<0:VDU3,4:PRINT:VDU5
245 IFk%=1PROCZ
370 VDU3,4:PRINT"PRESS RETURN  ";:INPUT""A:IFA<>1GOTO95
375 REM
380 VDU4,7:PRINT"PRESS RETURN  ";:INPUT""A:MODE7:PRINT'''"PRESS f2 TO RUN AGAIN"'':*KEY2CLEAR:GOTO5;M
385 END
390 DEFPROCM
395 FORA%=0TON%-1:E%(A%)=INT((E%(A%)-I%)/S):N%(A%)=INT((N%(A%)-K%)/S):NEXT
400 FORA%=0TO40:FORB%=0TO40:D=0:FORC%=0TON%-1
405 Q=1/SQR((B%-E%(C%)-.5)^2+(A%-N%(C%)-.5)^2):IFQ>1D=D+1 ELSED=D+Q
```

```
410NEXTC%:X%(B%,A%)=1000*N%/D:VDU4:PRINT'';A%;'',"B%;:VDU5:NEXTB%:NEXTA%
415FORC%=0TON%-1:B%=E%(C%)+5:A%=N%(C%)+5:PROCI:D%(C%)=W%:NEXT
420ENDPROC
425DEFPROCJ
430PROCI:IF(E%<=M%ANDF%<=M%ANDG%<=M%ANDH%<=M%)T=T+1:ENDPROC
435IF(E%>=M%ANDF%>=M%ANDG%>=M%ANDH%>=M%)ENDPROC
440PROCQ:IFW%<M%:T=T+E  ELSEIFW%>M%:T=T+(1-E)ELSET=T+1/2
445ENDPROC
450DEFPROCQ
455E=0:Q=0:U=E%:V=W%:PROCF:IFQ=1u=E%/2:v=F/2:PROCK:Q=X:P=Y
460IFQ>0MOVEQ,P:G=F/2:U=E%:V=H%:PROCF:IFQ=2u=0:v=F:PROCK:E=E+G*(F-G)/2:PROCX:DRAWX,Y
465IFQ>0Q=3:MOVEQ,P:U=E%:V=F%:PROCF:IFQ=4u=E+G*(F-G)/2:PROCK:DRAWX,Y:Q=0
470U=F%:V=W%:PROCF:IFQ=1ORQ=3u=F%:v=0:PROCK:E=E+F^2/4:PROCX:0=X:P=Y:IFQ=4:E=E+(1-G-F)*(G+F)/2:DRAWX,Y
475IFQ=3Q=0
480IFQ>0MOVEQ,P:G=F/2:U=F%:V=E%:PROCF:IFQ=1u=F:v=0:PROCK:E=E+G*(F-G)/2:PROCX:DRAWX,Y
485IFQ>0Q=3:MOVEQ,P:U=F%:V=G%:PROCF:IFQ=4u=1v=F:PROCK:E=E+G*(F-G)/2:PROCX:DRAWX,Y:Q=0
490U=G%:V=W%:PROCF:IFQ=1ORQ=4u=1-F/2:v=1-F/2:PROCK:E=E+F^2/4:PROCX:0=X:P=Y:IFQ=4:E=E+(1-G-F)*(G+F)/2:DRAWX,Y
495IFQ=3Q=0
500IFQ>0MOVEQ,P:G=F/2:U=G%:V=F%:PROCF:IFQ=2u=G%:v=F%:PROCK:E=E+G*(F-G)/2:PROCX:DRAWX,Y
505IFQ>0Q=3:MOVEQ,P:U=G%:V=H%:PROCF:IFQ=4u=1-F:v=1:PROCK:E=E+G*(F-G)/2:PROCX:DRAWX,Y:Q=0
510U=H%:V=W%:PROCF:IFQ=1ORQ=4u=F/2:v=1-F/2:PROCK:E=E+F^2/4:PROCX:0=X:P=Y:IFQ=4:E=E+(1-G-F)*(G+F)/2:DRAWX,Y
515IFQ=3Q=0
520IFQ>0MOVEQ,P:G=F/2:U=H%:V=G%:PROCF:IFQ=2u=H%:v=G%:PROCK:E=E+G*(F-G)/2:PROCX:DRAWX,Y
525IFQ>0Q=3:MOVEQ,P:U=H%:V=E%:PROCF:IFQ=4u=0:v=1-F:PROCK:E=E+G*(F-G)/2:PROCX:DRAWX,Y:Q=0
530U=E%:V=W%:PROCF:IFQ=4u=F/2:v=F/2:PROCK:E=E+(1-G-F)*(G+F)/2:PROCX:DRAWX,Y
535ENDPROC
540DEFPROCF
545IF(U<=M%ANDV>=M%)OR(U>=M%ANDV<=M%)F=ABS(U-M%)/ABS(U-V):Q=Q+1
550ENDPROC
555DEFPROCK
560X=I%+S*(B%-5+u):Y=K%+S*(A%-5+v)
565ENDPROC
570DEFPROCI
575E%=X%(B%,A%):F%=X%(B%+1,A%):G%=X%(B%,A%+1):H%=X%(B%+1,A%+1):W%=(E%+F%+G%+H%)/4
580ENDPROC
615DEFPROCR
620REPEAT:G%=0:FORA%=0TON%-2:E%=D%(A%):F%=D%(A%+1):Q%=E%(A%):T%=N%(A%):U%=E%(A%+1):V%=N%(A%+1)
625H%=E%-F%:IFH%>0:E%(A%)=U%:N%(A%)=Q%:N%(A%+1)=T%:D%(A%)=F%:D%(A%+1)=E%:G%=G%+1
630NEXT:UNTILG%=0
635ENDPROC
665DEFPROCZ
```

Program 3 *Contd.*

```
670IFp%=1VDU2
675S$="Isopleth areas":i%=73
680E=V(O)/100:@%=&20209:MOVE0,930:PRINT'S$'"max = ";V(O)" ha":VDU3:@%=&20110
685IFE=0GOTO710ELSEMOVE0,810:PRINT"% ha (%max) "
690FORA%=1TO16:PRINTSTR$(100-A%*5);" "V(A%)/E")   ":NEXT
695MOVE1270,45:DRAW600,45:MOVE600,455:PRINT;"100":MOVE500,255:PRINT;" 50"
700MOVE500,55:PRINT;"  0":FORA%=0TO4:MOVE(600+(A%*150),25:PRINTSTR$(100-(20*A%)):NEXT
705MOVE520,355:PRINT"%A":FORA%=0TO16:MOVE600+(38*A%),55+(V(A%)*4/E):VDU(i%):NEXT
710IFE=0GOTO725ELSEIFp%=1VDU2
715MOVE0,2000:FORA%=0TO3:FORB%=1TO4:PRINTSTR$((20-B%-4*A%)*5)"%=";V(B%+4*A%)"ha.("V(B%+4*A%)/E")   ";
720NEXTB%:PRINT:NEXTA%:VDU3
725ENDPROC
760DEFPROCP
765I%=10^8:J%=0:K%=10^8:L%=0:FORA%=0TON%-1:IFE%(A%)<I%I%=E%(A%)
770IFE%(A%)>J%J%=E%(A%)
775IFN%(A%)<K%K%=N%(A%)
780IFN%(A%)>L%L%=N%(A%)
785NEXT:w%=I%:e%=J%:s%=K%:n%=L%:IFk%=0o=500:p=300 ELSEo=700:p=500
790J%=J%-I%:L%=L%-K%:IFJ%>L%S=J%/30 ELSES=L%/30
795e=(1100-o)/(S*30):n=(900-p)/(S*30)
800ENDPROC
805DEFPROCX
810IFk%=0OR(k%=1AND(q%=0ORh%=z%))X=(X-w%)*e+o+12:Y=(Y-s%)*n+p-15  ELSEY=2000
815ENDPROC
940DEFPROCY
945VDU4:PRINT"FINDING MEAN FIX";:VDU5
950A=0:FORA%=0TON%-1:B=0:FORB%=0TON%-1:IFA%=B%GOTO960
955IFE%(A%)=E%(B%)ANDN%(A%)=N%(B%)B=B+1 ELSEB=B+1/(SQR((E%(A%)-E%(B%))^2+(N%(A%)-N%(B%))^2))
960NEXTB%:IFB>A:Q%=A:A=B
965NEXTA%:MOVE(E%(Q%)-w%)*e+o,(N%(Q%)-s%)*n+p:VDU143
970ENDPROC
1000DATAM9-9/85,41,46,48,45,38,47,40,44,45,38,47,42,43,41,44,45,38
1010DATA48,46,49,48,49,44,44,44,51,40,49,45,51,39,47,48,49,40,45,50,49,45,42,51,44,51
1020DATA47,47,43,47,42,45,47,44,42,45,42,44,57,47,50,48,42,50,-1,-1
1050DATAM7-9/85,-999,-999,48,45,41,44,36,47,48,49,48,50,29,49,51,48,30,52,41,43
1060DATA48,48,60,47,38,47,48,49,53,47,41,47,44,47,48,49,43,49,50,49,39,53
1070DATA41,51,50,48,40,50,40,47,41,46,36,51,40,47,40,47,35,51,-1,-1
1100DATAAF8-9/85,15,31,16,18,20,24,14,29,14,34,19,39,17,34,14,28,14,29
1110DATA19,38,15,27,18,34,17,34,15,31,17,31,20,41,15,29,15,31,28,34,21,27
```

1120 DATA 27,30,32,37,24,39,26,29,27,38,15,31,15,31,29,34,16,30,15,28,-1,-1
1150 DATAAF2-9/85,35,34,43,32,34,32,34,32,52,48,35,36,31,33,26,41,37,39,29
1160 DATA 51,48,35,31,35,31,46,45,50,49,43,37,41,31,39,33,43,31,43,32,37,33
1170 DATA 37,33,61,47,34,33,36,33,62,47,39,31,34,32,35,31,35,32,-1,-1
1200 DATAAF2-9/85,27,48,37,37,25,47,25,48,40,45,38,47,42,43,28,47,25,46,46,47
1210 DATA 26,46,28,44,27,47,28,43,26,45,20,40,29,43,30,39,26,46,41,50,24,46
1220 DATA 34,43,34,43,30,46,26,43,28,49,27,48,31,45,27,47,27,47,28,47,-1,-1
1250 DATA -1

Index

Acepromazine, 102
Acoustic tagging, 4
Active array antenna, 152
Activity centres in range analysis
  core convex polygon model, 174
  harmonic mean model, 172
  parametric models, 171
Adcock antenna, 17
  bearings with, 122
Adhesive, *see* Glue-on tags
Adverse affect avoidance, 99–101
  long-term behaviour, 99–101
  short-term shock, 99
Aircraft tracking, 136–143
  antennas, 136–137, 143
  for hawk dispersal, 139–140
  and lost tags, 133
  receiver programming, 140
  search pattern, 137–139
  signals and tag position, 140–142
  and height, 142–143
Amphibians, 38, 104
Anaesthetics
  in collars, 46
  darts, 3, 46
    radio location, 101
  for implantation, 109–110
  sedation, 101–102
Analysis techniques, 163–178
  density, 163–165
    grid trapping, 164–165
  survival, 165–167
    and signal loss, 165–166
    partial mortality, 166–167
    Cox proportional hazards model, 167
  event records, 167–168
  distance and speed records, 168
  range analysis
    grid cells, 168–170
    outline techniques, 170–178
Antennas, receiving, *see also*
    Antennas, transmitting
  and frequency, 10–11
  portable, 15–18
  on vehicles, 18–21
    retractable, 134–136
    fixed, 21–22
  on aircraft, 136–137, 143
  orientation, 117–118
Antennas, transmitting, 5, 31–33
  for fish, 98
  fitting on tags, 86–97
  loop, 17, 32–33, 61, 72–77
    bearings with, 118, 119, 122
  subcutaneous, 110
  in water, 3–4
  and water entry, 81
  whip, 70–72
Assembly of tags, 67–98
  antennas, 70–77
    loop, 72–77
    whip, 70–72
  circuit
    cells, 78–79
    labels, 83
    potting, 80–83
    switches, 79–80
    testing, 77–78
  power sources, 67–70
    cell lead attachment, 67–69
    diode isolation, 69
    solar panels, 69–70
  types
    1–2 g, 83–85, 91–92
    2.5 g implant, 96–97
    3 g fish, 97–98
    12–14 g, 85–90
    25–125 g, 92–96
Attachment of tags, 99–113
  adjustment and detachment, 111–113
  adverse effect avoidance, 99–101
  sedation for, 101–102
  techniques, 102–111

Back-bearings, 16, 132
Back pack assembly, 85–87
Badgers, 105, 144

Bats, 38
Bearings, 118–123
Bears and expandable collars, 112
Beavers, implantation, 110
Behaviour
  prediction from movement tracking, 131
  and tagging, 73, 79, 99–101
    squirrels, 109
    game birds, 103–104
Benzocaine, 110
Beta lights, 1, 145
Bias
  in capture, 3
  in surveillance observations, 144
Biodegradable sutures, 112–113
BNC connectors, 18
Boats, tracking from, 143
Bobcats and expandable collars, 112
Button cells, soldering, 67–69

Capacitors, 51
Capture, 3, *see also* Adverse effect avoidance; Grid trapping
  aids, 45–46
  shock, 3
  of hares, 99
Cells, 5, 28–30
  capacity, 29
  diode isolation, 69
  lead attachment, 67–69
  lithium/copper oxide, 78, 85–86
  shelf life, 30
  solar panels, 69–70
  zinc/air, 29–30, 78–79, 83–84
Chewing of tags, 79, 81
Cluster analysis of range, 175–178
  versus harmonic mean model, 177–178
Collars, 35–37
  anaesthetic, 46
  and animal growth, 111–112
  assembly
    2 g, 91–92
    125 g, 94–96
  attachment, 108–109
Cost, 7–8, 13–15, 24
Cox proportional hazards model in survival estimate, 167

Crocodiles, 38
Crystals, 26–27, 50
  specificiation and testing, 49
Cyanoacrylate glue, 82

Darts, anaesthetic, 3, 46, 101
Data loggers in telemetry, 160–161
  and mortality sensing, 161
Data
  storage and tag design, 43
  recording
    accuracy, 146–148
    mortality *versus* tag failure, 148–150
    position sampling, 145–146
    radio surveillance, 143–145
  requirements, 8
Density estimates, 163–165
  grid trapping, 164–165
Dental acrylic for potting, 81
Diffraction, 115–116
Diodes
  in cell isolation, 69
  and solar charging, 69–70
Dipole antennas, 16
  in active arrays, 152
  bearings with, 122–123
Direction finding, fixed stations, 22
Disappearances, *see* Lost tags
Dolphin harness, 113
Dry joints, 77–78
Ducks, implantation and anaesthetics, 110

Eagles, bald, satellite tracking, 45
Epoxy
  conducting, for lead attachment, 68
  for gluing, 102
  for potting, 48, 51, 58–60, 80–87, 94
Equipment, 9–46
  antennas, 15–22
  capture aids, 45–46
  frequency, 9–11
    and antennas, 10–11
  receivers, 11–15
  satellite tracking, 44–45
  tag design, 28–44
    antenna format, 31–33

Equipment – *continued*
   cells, 28–30
   data storage, 43
   mounting, 33–39
   on-off options, 43–44
   sensors, 39–43
   solar power, 30–31
   transmitters, 22–28
Etorphine, 101

Ferrite core antennas, 32, 72
Field techniques, 2
Fish tagging, 97–98
   aircraft location, 143
   and frequency, 11
   mounting of tags, 38–39
   presence/absence recording, 156
   suturing, 110
      biodegradable, 112–113
Fixed stations, 151–161
   and direction finding, 22
   intermittent sampling, 158–159
   presence/absence recording, 156–159
   radio location, 151–156
      automatic, 151–154
      manned, 154–156
   radio telemetry, 159–161
Flux-cooled soldering of cells, 67
Foxes radio surveillance, 144
Frequencies, 9–11
   and antennas, 10–11
   competition for, 9–10
   drift and temperature, 12, 14
   separation, 12
Frogs, 38

Galliformes, necklace for, 89–90
Glue-on tag, 37–38, 102
   assembly, 83–85
Goshawk, *see* Hawks
Grant Squirrel Logger, 160
Grid cells, 168–170
Grid trapping in density estimation, 164–165
Grooming and antenna damage, 73
Grouse, blue and tag weight, 100
Gulls, backpack mounting, 33

Hares
   capture shock, 99
   tagging, 36
Harmonic mean model in range analysis, 172–173
   versus cluster analysis, 177–178
Harnesses, 103–105
   for birds, 103–104
Hawks,
   density estimate, 163–164
   dispersal, aircraft monitoring, 139–140
   hunting, and radio surveillance, 144
   posture sensor, 39–41
   and signal interpretation, 130–131
   tail tag mounting, 107
Heart beat tags, vii, viii, 43
Hedgehogs, 7, 38
Horizon effect and signal detection, 155–156
Hyperbolic system in radio location, 152–154

Imobilon, 101
Implantation
   of fish, 39
   of otters/seals, 37–38
   of snakes, 38
   tag assembly, 96, 97
      of fish tag, 97–98
   technique, 109–110
Ingested tags, 81, 110–111
   salmon, 111
Interference (radio), 115–116
Isoline range use contours, 172–173

Ketamine hydrochloride, 101

Latching circuits, 44
Leg mounting, 34
Licence for transmitter building, 8
Lidocaine hydrochloride, 110
Lincoln-Seber Index for density of elusive species, 164
Lithium
   cells, 28, 29, 78
   fitting, 85–89

Lizards, 38, 104
Loop antennas, 17, 32–33, 61, 72–77
  bearings with, 118, 119, 122
Lost tags, 128–130, 132–133

Marine mammals and data storage, 43
Marking behaviour, 42
Martens, tag loss, 37
Mercury cells, 28, 67, 83
Methoxyfluorane, 101
Mink, 37
Mortality, see also Survival estimates
  and Cox proportional hazards
    model, 167
  sensing, 41, 131
    data loggers, 161
  versus tag failure, 148–150
Motorised tracking, 134–143
  aircraft, 136–143
    antennas, 136–137, 143
    search pattern, 137–139
  boats, 143
  road vehicles, 134–136
Mounting of tags
  birds, 33–35
  fish, 38–39
  mammals, 35–39
  reptiles, 38
Movement
  cues, 130–131
  sensing tags, 41
  tracking, 131–132

Nasal saddle tags, 35
Necklace tags
  assembly, 89–90
  attachment, 107–108
  on birds, 35
  on mammals, 36
NiCad cells and solar recharging, 31

Otters, glued tag mounting, 37–38
Owl harness, 103

Parametric models in range analysis, 171–172
Passerines, 33, 89

Patagial tags, 35
Petrels, giant, satellite tracking, 45
Pheasant survival, 100
Phencyclidine hydrochloride, 102
2-Phenoxyethanol, 110
Phosphoric acid flux, 67
Poncho tags, 34–35
Posture sensing tag, 39–41
  assembly, 87–89
  circuitry, 61
  signals from 130–131
Potting, 80–83
  dental acrylic, 81
  epoxy, 48, 41, 58–60, 80–87, 94
  moulds, 82
  and tuning, 72–73
Power sources, see Cells,
Predators/prey, 3, see also Raptors
  radio surveillance, 144
Preliminary considerations, 1–8
  alternatives compared, 1–2
  data requirements, 8
  expense, 7–8
  tag characteristics versus animal
    type, 3–7
  time, 2–3
Presence/absence recording, 156–158
  intermittent sampling, 158
Probabilistic circles model for ranges, 171–172
Programs for range analysis, 199–215
  for cluster analysis, 208–211
    procedures, 202
  data and displays, 199–200
  harmonic mean contouring, 212–215
    procedures, 202–203
  for parametric model, 204–207
    procedures, 200–201

Rabbits, thermistor tag for, 94
Radio
  location, fixed stations,
    automatic, 151–154
    manned, 154–156
  surveillance, 143–145
    and disturbance, 144
    and observational bias, 144
Radioisotope tracking, 2
Rain forests and frequency, 11

Index

Range (distance), 5–6
  and antennas, 11
  in water, 3–4
Range (territorial) analysis, 168–178
  cluster analysis, 175–178
  core convex polygons, 173–175
  grid cells, 168–170
  harmonic mean model, 172–173
  parametric models, 171–172
  programs, 199–215
    for cluster analysis, 208–211
    harmonic mean contouring, 212–215
    for parametric model, 204–207
    procedures, 200–203
Raptors, 87–89, 106–107, *see also* by name
Rasmussen Index in grid cell data analysis, 169
Receivers, 11–15
  programmable, 14, 15
  project size, 11–14
Reed switch, 44, 79–80
Rejection of young, 99
Reptile tagging, 38
  temperature recording, 159
Resistors, 51

Salmon, ingested tags, 111
Sampling saturation in position fixes, 146
Satellite tracking, 9, 44–45
Sea-otters, implantation, 110
Sea trout ingested tag retention, 111
Seabirds, *see also* by name
  diving and data storage, 43
Seals
  glue-on tags, 37–38
  harnesses, 105
  presence/absence recording, 156
Security, 159–160
Sedation, 101–102
Signals
  absorption, 115
  and cable length, 21
  diffraction, 115–116
  horizon effect, 155–156
  and movement, 130
  polarisation, 115, 118, 126

Silicone rubber
  potting, 81
  sealing of necklace, 90
Silver cells, 28–30, 67, 83
Snake tagging, 38
  ingested tag retention, 111
Solar power, 30–31, 69–70
Soldering technique, 67–68
  dry joints, 77–78
Sonar, 4
Squirrel thermistor tag, 92–94
Statistical tests and position sampling, 145–146
  see also Range (territorial) analysis
Strain relief, antenna exit, 71–72, 74
"Superglue", 82
Survival estimates, 165–167
  confidence limits, 166
  Cox proportional hazards model, 167
  and partial mortality, 166–167
  and signal loss, 165–166
Sutures
  biodegradable, 112–113
  for fish, 110
Switches, 79–80
  reed, 44, 79–80
  tilt, *see* Posture sensing tag

Tail mounting, 33–34
  assembly, 87–89
  attachment, 105–107
Telemetry, 159–161
  data loggers, 160–161
Temperature
  and frequency drift, 12, 14
  recording, 159
  sensing tags, 41–42
Thermistor, 41–42, 50
  collar assembly, 92–94
  in construction, 61
Tilt switch, *see* Posture Sensing tag
Toads, 38
Tortoises, 38
Tracking, 115–150
  bearings, 118–123
  data recording
    accuracy, 146–148
    mortality, 148–150

Tracking – *continued*
   position sampling, 145–146
   radio surveillance, 143–145
  distance/position, 123–125
  fix, 125–156
  "lost" tags, 128–130, 132–133
  motorised tracking
   aircraft, 136–143
   boats, 143
   road vehicles, 134–136
  setting up, 116–118
  signals from animals, 130–133
   disappearances, 132–133
   movement cues, 130–131
   movement tracking, 131–132
  triangulation, 126–128
  wave propagation principles, 115–116
Transistors, 50
Transmitter circuits, 22–28
  buying, 25–28
  and crystal harmonics, 26–27
  one/two stage and range *versus* weight, 24–25
Transmitters, building, 47–66
  antennas, loop, 61
  circuits, single stage, 49–50
  components
   single-stage, 50–51
   two-stage, 62–63
  construction
   single-stage, 51–56
   two-stage, 62–66
  frequencies, alternative, 60
  potting, 58
  sensor circuitry, 61
  tools, 48–49
  troubleshooting, 57–58
  tuning
   single-stage, 58
   two-stage, 66
Transponders, 2

Trapping, grid, 164–165
Triangulation, 126–128
Tricaine methyl sulphonate, 110
Tuning, 58, 66
  after potting, 72–73
  with zinc/air cells, 78–79
Turtles, 38
  harness, 113
Type Approval, 26–27

UHF
  connectors, 18
  tags, 130

Vehicle tracking, antennas, 18–21, 134–136
Visual markers, 1, 145

Water
  conductivity, 3–4
  proofing, 81, 98
Weight, 5
Whip antennas, 70–72
Wire for antennas, 70–71
Woodcock, 113
Woodland and signal diffraction, 115–116

Yagi antennas, 10, 15–17
  on vehicles, 18–21
   retractable, 134–135
  in fixed stations, 21–22
  bearings with, 118, 120–121
  fixes with, 125–126

Zinc/air cells, 29–30, 78–79 83–84,
Zinc/manganese dioxide cells, 28